The
OTTERS'
TALE

SIMON
COOPER

WILLIAM
COLLINS

To Mary and Minnie – muses both.

'It'll be all right, my fine fellow,' said the Otter (to Ratty). 'I'm coming along with you, and I know every path blindfold; and if there's a head that needs to be punched, you can confidently rely upon me to punch it.'

Kenneth Grahame, *The Wind in the Willows*

William Collins
An imprint of HarperCollins*Publishers*
1 London Bridge Street
London SE1 9GF

WilliamCollinsBooks.com

First published in the United Kingdom by William Collins in 2017

22 21 20 19 18 17
11 10 9 8 7 6 5 4 3 2 1

A catalogue record for this book is
available from the British Library.

ISBN 978-0-00-818971-6

Typeset by Palimpsest Book Production Ltd, Falkirk, Stirlingshire
Printed and bound in Great Britain by Clays Ltd, St Ives plc.

MIX
Paper from
responsible sources
FSC™ C007454

FSC™ is a non-profit international organisation established
to promote the responsible management of the world's forests.
Products carrying the FSC label are independently certified to assure
consumers that they come from forests that are managed to meet the
social, economic and ecological needs of present and future
generations, and other controlled sources.

Find out more about HarperCollins and the environment at
www.harpercollins.co.uk/green

CONTENTS

Prologue   1

Chapter 1   Alone and afraid   5

Chapter 2   A place to call home   27

Chapter 3   Something in the air   37

Chapter 4   Alone but determined   65

Chapter 5   And then there were five   81

Chapter 6   And then there were four ...   95

Chapter 7   A country playground   117

Chapter 8   Life in the food chain   133

Chapter 9   To Autumn   153

Chapter 10   It's a dangerous life   171

Chapter 11   Solstice   187

Chapter 12   Lutran's tale   209

Chapter 13   Coming of age   233

Chapter 14   The long journey   247

Epilogue   261

Bibliography   263

Acknowledgements   265

Index   267

PROLOGUE

Looking back on it, I was something of a fool; the signs had been there for years but it took a fall of January snow finally to reveal what I should have known all along. As night turned into day, the virgin snow around the lake was anything but virgin, the peninsula that divided the lake from the river criss-crossed with seemingly a thousand footprints or more. From river to lake, lake to river and back again, the night-time visitors had clearly been busy, the five clawed paw prints exposing the green grass beneath the broken snow. This animal runway was as churned up as any busy city-centre pavement, but with particularities that told its own unique tale.

The river bank spoke of great effort, the snow ground into mud. Deep impressions in the turf atop the bank were clearly purchase points, the lower bank a mess of icy earth where the creatures had scrabbled to haul themselves from the water. Where the track marks met the lake was a different story. An icy slide, which looked as fun as that of any water park, was worn smooth with regular use, forming the connection between land and water. At the approach a patch of snow, maybe the size of a large door mat, was crushed flat – smooth evidence, to my mind at least, of someone or something lying and rolling in the snow.

In the dull light of pre-dawn one corner of battered snow beneath the tall alder caught my eye. It looked different to the rest and, sure enough, as I approached I could see the mottled snow was flecked with blood, with a bright red patch at its centre. Little bright silvery grey specks, at first unfamiliar, decorated this collage of nature. I stooped down, licked my finger and dabbed at one. A fish scale, shining like translucent mother-of-pearl, glinted back at me. The cogs in my head were gradually clicking into alignment.

A raspy, wheezy cough cut through the silence, and there, at the base of the alder, on the roots that formed a sinewy platform at the lake edge, sat the otter that I would one day know as Kuschta. In truth, she seemed calmer about our accidental meeting than I was. In that fraction of a second in which our eyes locked she assessed me, dismissed me as irrelevant and then turned, in one fluid movement pouring herself into the lake. I, on the other hand, stood rooted to the spot, uncertain what to say or do. I mean really, how daft is that – what could you ever say to an otter? Or do? Well, I did nothing. She, clearly the more evolved one in this particular situation, surfaced a few yards out from the bank before heading for the island that sits in the middle of the lake. On reaching its edge she emitted a single *eek*, which was echoed a moment later by a short symphony of *eeks* that soon took form as four dark shapes plopped from the island into the water to join her.

Prologue

From my rooted spot I could easily track the progress of the swimming party across the lake as they set course for the outflow where it joined the river at the waterfall created by the weir. The five flat, domed heads glistened against the inky blackness of the water. They were hurried rather than panicked, with the young otters swimming in a rough V formation behind their mother. As they scrambled over the weir I lost sight of each in turn, but it was a long time before I knew they were completely gone. For a while as they headed downstream I could hear them cavorting and splashing as they went, *eeking* to each other every few seconds in that otterly way that says, 'Don't worry, I'm fine, I'm still with you.'

But eventually all I had was silence and the red sky of dawn. Somewhere downstream, in the water meadows and woods that border the river, the otters would seek refuge from the day, curling up in the warm, dry comfort of a rotten tree trunk until dark. I'd lost them for now but somehow I knew they would return.

CHAPTER I

ALONE AND AFRAID

Two years earlier

As dusk started to fall, Kuschta gradually uncoiled her body, stretching away the stiffness of a day spent asleep. Sniffing the air, she could tell the holt was empty without even opening her eyes. There was nothing unusual in that, but she was comforted by the slight warmth radiating from the indentation left by her mother in the soft bed of the rotten willow trunk which they shared. She was clearly not long gone. Kuschta weighed up her options. She guessed her mother would not be far, probably down at the weir pool diving for eels – easy pickings, as they gathered in great numbers before their summer migration to the sea.

In truth, Kuschta had options, but only of the no-win kind. On that fateful evening the outcome was to be the same, whatever her decision. Whether she stayed in the holt or went in search of her mother she was destined to greet the following dawn alone. But knowing none of that, Kuschta assumed her mother would return as she had done every day of her fourteen-month life. Maybe with a tasty eel that they would share? So oblivious to the future, Kuschta chose to stay put, recoiled herself, buried her nose in the crook of her hind leg, closed her eyes and was instantly asleep.

You could walk close by the spot where Kuschta slept and not give it a second glance. In fact, I'd hazard that even if you stopped and stared you might not be much the wiser. Otters are not like badgers, which dig elaborate setts, creating multiple cave-like entrances with the spoil of their digging spread around for all to see. In their choice of home otters are pragmatists, moving between 'holts', which are usually tunnels amidst the roots of trees beside the river, and 'couches', well-concealed resting places above ground. Neither is particularly easy to spot because they are so much a part of the landscape, used by generation after generation of otters. Twenty, thirty, forty years of continuous use is not uncommon – even a century has been recorded. This we know from otter hunts that combed every inch of bank until they were banned in the 1970s. A diligent huntsman would know every hover, as some called them, returning time and again to seek out their prey and record the locations for future hunts.

Otters are not builders like, say, beavers; they take what they find and adapt it. The best holts are created by Mother Nature. An ash or a sycamore grows tall beside the river until gravity takes a hold as the water erodes the bank beneath, so the tree starts to lean out over the river. These trees are well adapted to such a pose, the wide shallow roots providing enough support for some considerable elevations of lean. But it is in that process of tipping that the den is made; the movement lifts the bank to create a cavity under the roots, which in turn becomes the canopy of the holt. And in the otters go. Some judicious digging will create a labyrinth of dry tunnels and, if all goes according to plan, there may even be an underwater connection to the river.

Couches, on the other hand, are rather more at the making of otters, but they are, as the name suggests, the more informal of the two habitats. A pile of reeds, dry moss or leaves in a thicket of brambles a few yards from the river would be typical. It's more of a good weather than a bad weather sort of place, though not always. In wet flood plains where dry holts are scarce, elaborate couches are created as alternative homes, but more generally the couch is the resting place where otters feel safe to sleep, catch the sun and play whilst whiling away the daylight hours, hidden from view.

I guess Kuschta would care little for my subtle differentiation between couches and holts. All she knew was that the hollowed-out crack in the willow branch had long been a favoured resting place for the family; warm, inviting and familiar. The willows that thrived by the river had this strange way of growing that helped the otters; shooting up fast, they soon outgrow themselves so much that the limbs burst or crack open lengthways before snapping away from the trunk and falling to the ground. Laid out, the cracked branches look a little like an open pea pod, and at first there would probably only be just enough room for an otter to squeeze inside the fissure in the wood. But in time the timber would start to rot from the inside out, the constant comings and goings of the otters gradually hollowing out the trunk. The bark and cortex, still connected to the mother tree, stay alive even to the extent that the bright green shoots continue to grow up to create a sort of curtain in front of the hollow. It is as natural a hiding place as you'd ever find.

It was well into the night when Kuschta woke up with a start; something or someone was passing by. She had no reason to be scared, but she was, freezing rigid until the sound faded into the distance before she raised her head to check the hollow. Empty. She pawed at the soft, rotten wood where her mother usually sat. Cold. Cold as if she had never been there. Something wasn't right. Wishing it wasn't so, Kuschta stared out at the river for a little while, the landscape bathed in the silver light of a half moon, until she reached a decision – she'd go to find her mother.

Parting the willow-whip curtain, Kuschta pushed herself out of the hollow and slid into the water. In her mind there was no doubt she would find her mother at the weir pool. It was one of their favourite haunts. As she swam she became more confident in her decision. Familiar landmarks marked the route. The moon lit the way. The weir was not far. Turning the last bend with it just ahead, Kuschta slowed her pace. She half expected to see the silhouette of her mother on the wooden beam that braced the right-hand side of the structure. The two of them often sat there to share the spoils, but tonight it was empty. No matter. Kuschta stopped paddling, letting the current take her along whilst straining her ears for familiar sounds above the regular pounding of the water as it crashed over the weir. Nothing.

Scrambling up, she took in the whole pool from the vantage point of the beam with one swift movement of her head. The fast plumes of water that washed to the centre of the pool then gathered together to push out and on through the mouth to continue on as one river. The gentle slope of the grassy bank that led down to the water on the far side. The line of alders on the nearside, all gaunt and black against the night sky. All utterly familiar but totally absent of the one thing she sought. Confused and deflated, Kuschta settled down on the beam to wait for her mother to return. Time was her only hope.

It is probably better that at this point Kuschta doesn't know what we know, namely that she has been deserted by her mother forever. Deserted is a harsh word, but from such swift and brutal decisions, good will come. It's just that sometimes it doesn't seem that way at the time.

Otters are relatively unusual in the mammal kingdom of Britain, breaking with the norms of reproduction. In the standard way of things, offspring are born in spring, raised in the summer and go their own way by the autumn at the latest. But for otters it is somewhat different, the reproductive cycle being closer to biennial than annual, with the pups, the females in particular, staying with the mother until they are well over a year old. With so much time together, the bond between mother and pups is intense; the father is rarely a factor, moving on soon after mating. Otter pups are truly dependent on the mother; they can't swim or hunt without being taught. But even when armed with those basics they can't go it alone. Three months, six months, nine months, a year – they will starve without maternal oversight all the way into young adulthood.

The family group is everything in the upbringing of an otter. The litter, anywhere from one to four pups, will spend every waking and sleeping hour together with the mother until they go their separate ways. The first schism comes earlier for male pups. Approaching one year of age, he will be as big as his mother, and with increasingly unruly behaviour and becoming more dominant than she would like, the mother takes a stand against her son and drives him away. Shorn of her guidance and protection, this adolescent faces an uncertain future. Travelling alone, he has to fight for his life both in terms of finding a territory to call his own and food to live on. Mature males will care little for his intrusion and other mothers will be fiercely protective of their patch. He will find it hard to rest; the itinerant life will drain his strength as he is constantly moved on. His hunting skills are still evolving; surviving day to day in unfamiliar places is a constant battle. There are plenty for whom the battle is too much, dying of exhaustion and starvation. His sisters will eventually reach a similar fork in their lives.

Survival, and the search for food to ensure survival, can also determine otters' habitat. Sometimes the choice of a coastal home is made of necessity, driven along to the mouth of a river in a search either for territory or for food. Otters are very 'linear' in their habits and outlook; they rarely stray far from water, preferring to travel great distances along watercourses rather than heading inland. The only common exception to this is the birthing holt, which, for reasons we will discover, is located well away from the water. So when an otter can't find a territory to call his or her own, or possibly when there is not enough food, he keeps on travelling until, as with any river, he reaches the sea. Saltwater, freshwater – it is all the same to an otter. Coastlines offer the same opportunities for raising a family. Let's face it, if your kind has been able to span so many continents and landmasses, you are going to be pretty adaptable.

It is, however, worth making a distinction here between sea otters and otters that live by the sea. Kuschta – and every single otter that has ever existed in the British Isles – is of the family *Lutra lutra*, more commonly referred to as the

Eurasian or European otter, which is one of thirteen different known otter species around the globe. As the name suggests, the European otter is indigenous to Europe, but also to Asia and North Africa, with a truly phenomenal spread across the northern hemisphere. West to east, with a few exceptions, such as the Mediterranean islands, this species walks every landmass from the Atlantic Ocean to the Pacific Ocean. From northern Russia to the Indian Ocean, snowy Finland, dusty Morocco, the foothills of the Himalayas, humid Thailand or the frozen shores of the Bering Sea – if you know where to look, you'll find *Lutra lutra* in all these lands.

On the other hand, sea otters – *Enhydra lutris* – are an entirely different species altogether, found along the shallow coastal waters of Russia, California and Alaska in the northern Pacific Ocean. They rarely venture onto land, living entirely in the water, and are famous for floating belly up in the kelp forest that hugs the shoreline. Conversely, British otters that live by the sea are just your normal inland otters that have picked a different type of home. In the Scottish islands, where they thrive, this way of life is the practical alternative, where there is more productive coastline than freshwater river or loch.

As the sky started to show fire red behind the grey clouds in the pre-dawn, Kuschta knew it was time to move. She had seen and heard nothing during her night-time vigil. Neither friend nor foe had broken the silence or the surface of the eel pool. Stretching her stiff body on the beam, she knew she should have used the darkness to hunt, but with the sun coming up fast it was now too late. Darkness is the friend of otters; they navigate their world in the secret time between dusk and dawn. Daylight is the time for rest, night the time for travel, food and adventure; this much Kuschta had learnt from her mother. Impelled by the rising sun, Kuschta needed to make it back to safety or at least somewhere familiar. Sliding into the water, she swam quickly upstream, sending out unruly waves that rocked at the reeds on either bank, first startling a moorhen who let out an anguished squawk of protest, and then a water vole who simply paused his weaving amongst the stems until the commotion was past. Silhouetted against her skyline, the cattle grazed in the meadows, their scrunching remarkably noisy, even to her ears. A fox trotted across a bridge, no doubt heading, just as she was, to a daytime lair. The dawn chorus was reaching fever pitch as the birds laid territorial claim to a busy day ahead. The rural community of creature kind was resetting the natural order of things as the night shift headed for bed and the day shift repopulated a valley in all the pomp of its summer plumage.

Heading for the crack willow couch, Kuschta sensed her life had changed. She had no expectation of finding a familiar body settled on the spongy, rotten wood. She was alone, and that, for now at least, was the way it would be. Pushing herself head first through the willow screen, she settled into the couch. Hunger was now her greatest concern. For a while she stared out from her hide as the river slowly woke up. A trout sipped at insects caught in the surface film. Damselflies began their hovering dance. The kingfisher took up his perch ready to feed. Bees created a symphony of buzzing as they sucked the nectar from the buttercups that blanketed the water meadows. As the familiar sounds and the warmth of the morning overtook Kuschta, she slowly drifted towards sleep, but not before resolving to find food as soon as darkness came.

It is fitting that Kuschta spent her first nights alone in the hollow of a tree, as she took her name from the myth of Native American tribes who both revered and feared otters. In legend otters dwelt in the roots of trees, transforming themselves into human form at will. In some tribes the kuschta, which literally translates as 'root people', were friendly and kind, leading the lost or injured to safety. But for other tribes these shape-shifters did so for evil intent, to become the stealers of souls by guile, luring the naïve or unsuspecting away from home where they were transformed into otters, thus deprived of reincarnation and the consequent promise of everlasting life that was at the core of Native American belief. Interestingly, dogs were considered the best protection against these 'land otter people', being the only animals the kuschta feared as their barking would force them to reveal themselves or flee. Today dogs still have a similar effect on otters.

By the time dusk arrived Kuschta was ravenous, as hungry as she had ever been in her life. In all probability she had never gone this long without food, so just as soon as the darkness felt safe she pushed out into the river. The eel pool was the obvious destination, so she swam with purpose, confident in her ability to hunt down a meal solo; after all, she's been doing so for much of her life, but in the wild nothing is ever certain. Fish are faster than otters, which Kuschta had learnt very early on – over the first two yards a trout will always leave an otter trailing in its wake. Hours of fruitless pursuit had left her exhausted on the bank, reliant on the success of her mother or sharing with her siblings. But over time, through observation and imitation, she figured out that over ten yards, when stamina tells, the odds would start to tip in her favour. Add in a bit of stealth and suddenly she had a winning formula.

Cruising the pools, she'd use her super-sensitive whiskers to pick up the vibration of a fish. Arching her body, she'd dive head first, deep under the surface, preparing to hunt the fish from below. Sometimes if the moon was bright she'd see the outline of the fish above her, but more often it was her whiskers that were her guide. Propelling herself upwards with her webbed feet and powerful tail strokes, she'd accelerate towards her prey. For a moment she'd have the advantage of surprise, but fish are no slouches when it comes to sensing vibration; their lateral line, which runs the length of their body, is as good as any whisker sense and usually enough to grab that two-yard start. From then on it is ten seconds of life or death for the fish, success or failure for the otter. The two swim, leap, crash, weave and dive, creating mayhem in the pool as Kuschta tries to grab the fish in her mouth.

You might be tempted to think that the otter holds all the cards in this showdown, but in truth failure is the accepted outcome. What otters have is total determination; that instinct to try and try again. There are few hiding places for a good-size fish. In general they have to keep moving to survive. Movement equals vibration. Vibration equals discovery. When an otter has found a fish once, it will find it time and time again. Unsuccessful chases will be followed by more unsuccessful chases until, through tiredness or error on its part, a fish ends up clamped between those whiskery jaws or the otter accepts that the fish has won the day this time.

Approaching the eel pool, Kuschta aligned herself with one of the gaps in the weir that spouted water into the pool, knowing this to be the perfect cover for her attack. As she flopped over the lip she allowed the current to carry her out into the midst of the pool, the turbulence masking any evidence of her arrival. She really didn't have to expend much effort along the way, just use her tail and paws to course correct until the back eddy bought her to a gentle halt. In the still night dark she hung in the water, its pace now very slow as the champagne bubbles dissipated around her. Weightless and drifting, Kuschta was alert to the slightest movement – her whiskers, her eyes. A few yards off she sensed a fish coming her way, but before she could dive, it turned away. The fish were there, that much she knew. All she had to do was find one.

Calm to the task, Kuschta curved her body, using a slow tail beat to rotate in a wide circle, using eyes, ear and whiskers to scan the full circumference of the pool, the eddies, bubbles and swirls breaking the flat surface. Somewhere out there she sensed, rather than saw or heard, another fish slowly swimming away, unaware of her presence. Perfect. She arched her back, slid beneath the water, and when she was a few feet submerged she started to home in on the fish in a rising diagonal. As she gathered speed the distance between them narrowed. Advantage Kuschta. But not for long. The fish felt her coming as the bulk of her body pushed a sonic bow-wave ahead which reached him just before she did. That fraction of a second warning was enough to alert him to the danger. In a moment he went from languid to panic, his body squirting forward, flexing for speed as if electrified. Accelerating away, the gap widened, but that suited Kuschta just fine. The more the fish panicked the easier he was to track, as the chatter of vibrations came back to her through her whiskers. She hung on in his wake, letting her stamina blunt his speed. Across the pool they went, the distance narrowing with each yard as they headed for the far bank. For the fish the bank equalled safety, a chance to throw off his pursuer amongst the roots and undercuts. Kuschta knew this and put on extra speed to close the gap, getting herself close enough to lunge at the fish. In that final effort she bunched her body, then exploded forward, but just as her nose brushed the flank of the fish it twisted away, her jaws closing on nothing but water.

Kuschta surfaced for air, emitting a sharp cough as otters are apt to do after underwater effort – whether it is a reflex action, a clearing of the air passages or a prelude to a sharp intake of breath, I do not know, but it was to become one of the sounds that I will forever associate with otters. Head in the air and eyes shining bright, she readied for the next attack, wheezing as she gradually recovered her breath. She knew she only had to wait. All stirred up by the chase, the fish would be radiating vibrations, uncertain where to hide or swim. It would not take long for it, or maybe another fish equally disturbed by Kuschta's presence, to come back into her orbit. And sure enough, one came straight towards her at speed, whipping past and giving her only enough time to dive down at it. But her effort was more sound and fury than effect, as the fish was already out of danger by the time she had completed the lunge. Unconcerned, she paddled after the fish; sometimes she kept it close, other times it faded into the distance, until she picked up the tell-tale signs again. Around and across the pool they went. The pursued and the pursuer. For the trout it was about staying alive. For Kuschta it was about food. For both, in different ways, it was about survival.

Soon Kuschta sensed the fish was tiring, the sprints of flight slowing with each passing chase. Suddenly she felt the fish to her right, the two of them swimming parallel. This was her chance. One swift dive, turn and boom and she'd grabbed the fish by the soft underbelly. As the fish tensed and twitched she drove her teeth deep into its flesh, asserting her grip. She knew it wasn't the perfect hit, more to the tail than the head, allowing the trout to thrash and twist its body, but changing the grip of her jaws was no option. She'd learnt from bitter experience that was how meals escaped. So with the trout whipping about her head she swam for shore, digging her sharp claws into the bank to heave herself and the fish onto the grass. Subduing the fish by shaking it hard, Kuschta held it down with her front paws, let go of the belly and bit hard behind the head, extinguishing all life bar a few death twitches.

Otters don't gloat in victory, they simply get down to the job in hand, consuming the capture. Kuschta was too hungry to care anyway, tearing into the flesh, eating as much and as fast as she could. After twenty minutes, with the head and the best end of the fish safely in her belly, she paused to clean herself, licking away the smatterings of blood, scales and flesh. Silent and content, she settled down to rest for a while before she would finish the fish then head back to the rotten willow. But a noise in the distance changed all that. Up on the beam she knew so well, a figure appeared. An otter – bigger than any she had ever seen before. It first sniffed suspiciously at the ground where she had lain then tested the air. She froze as it held its head in her direction. For a moment she thought it was going to leap into the pool and head directly over to her, but it was distracted by another otter, more the size of Kuschta's mother, which came up by his side.

As the two nuzzled and groomed at each other's fur, Kuschta knew her time in the only place she had ever called home was over. The larger of the two was not an otter she recognised; the other may have been her mother but she could not be sure. Though part of her yearned to do it, she suspected, quite rightly, that no good would come of her revealing her presence. It was time to go. Edging away from the river, she headed for the woods and, keeping the sound of the water just within earshot, continued downstream for an hour or more. Moving on land is tiring for otters; yes, they can run quicker than you might think, with a gait not dissimilar to that of a greyhound, but given a choice, it is water at times of flight. Back in the river, Kuschta swam as fast as she could. Soon there were no more familiar landmarks, every fresh stroke taking her to a place she didn't know. In the space of two days and two nights she had lost her mother and her home. She was alone and afraid.

At dawn she could swim no further; her body was chilled to the core by too much time in the water. Dragging herself onto the bank, she shook herself like a dog, sniffed the ground and then padded through the long grass, occasionally stopping to sit up and look around. Soon she spied a dense clump of brambles not far from the edge of the river. Finding an opening, she pushed her way into the middle, the tendrils, laden with bullet-hard red blackberries still a month away from ripening, swinging closed behind her, keeping her safe from prying eyes and unexpected visitors. It was far from perfect, but for now, with the ground dry and the leaf mould soft, she gave into the sleep that her exhausted body craved.

Otters are not by choice nomadic, but in the months immediately after fleeing the eel pool Kuschta had few choices but to become so. Like all her breed, when fit and fed she was capable of covering great distances, but that was borne out of necessity, not choice. In her search for a place to call home, Kuschta found each successive territory occupied, and was forced to move on when her arrival became patently unwelcome.

That is the thing about otters. We tend to think of them as gregarious, social animals – the Disneyesque vision of a pellucid pool, ringed with trees, which is fed by a tumbling waterfall where the pups frolic and play whilst the parents keep an all-seeing eye as they sun themselves stretched out on the warm rocks. However, the truth is somewhat different; otters are really not very social animals. Once Kuschta had adapted to a life alone, it was the life she preferred. Of course she would join with another when the time for mating arrived, splitting immediately afterwards to become the dutiful single parent for as long as required, but once the pups were gone she'd return to the solitary life, the default choice for her species. Being non-social is all very well but it does require a space to claim as your own, and that was increasingly Kuschta's problem. Everywhere she went was occupied by people or otters.

A millennium of persecution has taught the otter a lot about people, not much of it good. Otters have learnt to be invisible, shunning the day and hugging the night. Where the river took Kuschta through towns she just kept swimming by night, staying in the shadows, so that people were oblivious to her presence. Sometimes she was forced to hole up for the day, but it was never a problem; culverts, outflow pipes and all manner of structures were plenty good enough for a layover until darkness returned. At first this was all very unfamiliar, but in her travels Kuschta soon learnt that she was, at least in respect of animals other than her own kind, top dog. The apex predator, as the biologists like to call species such as *Lutra lutra*. There might have been a time when wolves or bears roamed the British Isles that otters had something to fear, but today they firmly reside at the top of a food chain, upon which no other creatures prey. It is a pretty exalted place to be, but in every society – even within that of apex predators – a structure evolves with the weak at the bottom and the strong at the top. Kuschta, still a few months off physical and sexual maturity, was trying to find her place in that new order.

Otters are territorial creatures, but claiming homelands in a way that is really quite unusual, for despite all the attributes of a creature fit for fighting – lean, lithe, strong claws and sharp teeth – they choose another path. For aggression read avoidance. Apart from the rare occasion when two rutting males clash, they are the most anti-confrontational of animals, a trait which they achieve by the delicately phrased term of sprainting.

Spraints, to put none too fine a point on it, are defecations – markers set out along the river bank telling of who 'owns' which territory. If you are an otter you can tell a lot about your fellows from a simple sniff – age, sex, status, fertility – they are all there in the musky scent. Marking out your patch is a neverending task. Typically otters will cover anywhere from a quarter to a third of their territory on any given night, making thirty to forty deposits, far more than required by any digestive tract. Maybe this goes some way to explaining the origin of the word spraint, which comes from the French *épreindre*, which means 'to squeeze out'. The effort is neverending because the scent only lasts for just so long – a few days at most – requiring regular reinvigoration. Otters are creatures of habit when it comes to these marks, using the same place time and time again. Keen otter trackers will go to great lengths to find piles of dry guano, the remains of up to two hundred spraints, topped with a shiny new marker. Otter huntsmen were keen aficionados; they knew that otters returned to the same spots generation after generation. It is no surprise that these dropping piles were ideal for encouraging the hounds to pick up the scent. A huntsman would not be averse to a bit of scenting himself, squeezing the manure between his fingers to judge its age, along with a judicious sniff. Some hunters claimed similar nasal powers of identification to that of the otters themselves, announcing to the assembled hunt followers that they were on the trail of such and such an otter. You do wonder whether this was more about mystique than fact.

Kuschta became increasingly impatient as the new day wore on; night was a long time coming and she was hungry. In her temporary hide she pawed the ground for something to eat, but it was more of a distraction than a practical alternative – bugs and earthworms held little appeal. The previous night had been a hunting disaster; one small perch and an unlucky frog that she had stumbled across. As the sky started to darken she took her cue from the bats; if it was dark enough for them it was dark enough for her, so as soon as they began to flit across the blackening skyline, Kuschta was on the move. For days now she had been travelling ever upstream, keeping close to the river. Occasionally she had been driven inland by people, dogs or other otters, but essentially her path was that of the river bank. Sometimes she'd reach a confluence, the junction of the river offering a left or right choice with her taking the one seemingly least travelled.

Her progress was always halting, stopping to smell each spraint she came across, the new evidence to be considered before a decision could be made. Male or female was the first marker scent she'd evaluate. At first glance a male might spell bad news, but not necessarily so. A male otter will cover a huge territory, many times that of a female, so he expects to find a number of females within 'his' territory. On that basis she'd pose no surprise, interest or threat, being as yet still too young for mating. Females, on the other hand, were a different thing altogether. Though Kuschta offered no physical threat to a mother otter, who would be more than capable and willing of protecting herself and her pups, Kuschta would definitely be seen as taking a share of the neighbourhood food. It is interesting that otters moderate their reaction to competition for food very much according to the season. When times are good they are happy to share, and temporary visitors will be tolerated. When times are hard boundaries are, if not explicitly protected, to be more respected by an interloper.

The freshness or not of the spraint presented all sorts of conundrums; some good, some bad. If it was a fresh male mark, it signalled that he had been here and had moved on, and was unlikely to return for a few days. Stale? Well, he might be back sometime soon. Female spraints were different; lots of the same within close proximity told Kuschta that she was probably in the midst of a family territory – better to go. No female scents, or at least only ancient ones, were more complicated. Either she had stumbled across vacant territory, or maybe, if it was a productive area, the female was going through the secretive phase that bitches are apt to experience immediately before and after the birth of a litter, as they stop sprainting for a while. They change their normal behaviour to disguise their presence for the safety of the pups, which have to be left alone for a few hours each day whilst the mother hunts. Surprisingly, during this period the greatest threat to the litter comes not from other species but from otters themselves. Postmortems of roadkill dog otters regularly produce the remains of pups in the stomach. Now whether this is more often the father or another jealous male, nobody is entirely sure, but it surely happens and in all probability, as indicated by DNA matches, it is a bit of both. The fact is, this infanticide happens more often than we might like to believe. The reasons? Again, we don't really know, but one can best assume that the mothers would not have evolved such a protection strategy without good cause. For otters the genesis of single parenting may have many reasons.

But spraints are more than just markers of territory; they are, in twenty-first-century jargon, food management tools, used by otters in all sorts of ways. Firstly, the spraint can say 'I've just fished this pool, so give it some time to recover or we'll kill the goose that lays the golden eggs'. Or secondly 'Don't waste your time and effort. I've caught all there is to catch'. Conversely, a pile of spraints, the most recent a little old, tells the otter that historically this is a productive spot now ripe for hunting. And no spraints either mark virgin or barren territory – proceed as you choose. Likewise, the contents of the spraint, be it the remains of fish, eel, crayfish and so on, tell of what there was, and maybe there still is, to eat. Kuschta was still a few months off appreciating the final piece in this spraintology jigsaw, which tells of females ready to mate and males on the prowl. Her time for this would come soon enough; for now she was struggling to find that elusive home territory.

If you sat an otter down to discuss the whys and wherefores of territory, the first words out of its mouth would most likely be 'It's complicated'; for Kuschta and her kind the enforced itinerant months were a progression from ignorance to understanding. When she was young the home patch that her mother had carved out was uniquely theirs; as far as she was concerned, the otter world did not extend beyond her immediate family. If other otters, including her father, strayed close, her mother would ward them off long before they could approach the pups. But now, on the road, she had to negotiate her way through a competing world.

To the human eye it is nigh on impossible to tell where the territory of one otter starts and another ends. There are no clear dividing lines; otters are certainly no respecters of man-made boundaries and there is little in nature that will hinder their progress. Otters are legendary for the ability to cover enormous distances in a single night; twenty miles is well within the scope of most, but herein lies the problem – that twenty miles is along a river, inevitably passing through a multiplicity of other territories that rightfully belong to other otters. But our mustelids have created a social policy that some humans could do well to learn from. At the core is the family territory, which is sacrosanct. It is always accepted that other otters will 'pass through', but woe betide any that pause or, worse still, try to stay. There is no excuse for ignorance; those spraints are flags enough to say who lives where and why. Single female otters, on the other hand, are more relaxed; after all, they don't have pups to feed or defend so their territories are looser, overlapping at the edges. That said, they will have carved out a portion of the territory that they regard as 'theirs', expecting other otters to keep away. Males spread themselves much wider, their territory taking in maybe three, four or five females' homelands, which they continually traverse, taking as much as a week to cover the entire area. Of course, this assumes there is only one dog otter in any given super-territory. But nature would not allow that – replacements and competition are always required. There will be more than one male sharing the territory, a potentially ruinous situation with competition for land, females and food. But otters have evolved a daisy-chain hierarchy where each lesser male follows in the footsteps of his immediately superior male, organising his life so he precisely avoids meeting the others. Of course, there are clashes, violent fights and changes in the order. No otter chooses to be of lower caste, but, given a death or some other departure, there will be a shuffling of the pack.

It is a fascinating animal culture; despite being solitary creatures, otters as a breed have survived for 20 million years because, probably without realising it, they look out for each other. Avoidance saves wasteful territorial disputes. Designated family homelands allow pups to thrive. Complicated spraint markings spread out the population, avoiding overcrowding, preserving food stocks and limiting disease. Wandering males mix up the gene pool. For the perfect evolution you could not write a much better script – it is just a shame, as we will see later, that humans came into the midst of this with near-fatal consequences.

CHAPTER 2

A PLACE TO CALL HOME

Winter

As far as the weather is concerned, otters don't worry unduly about the seasons; they are perfectly adapted to anything the British climate might throw at them, unlike, say, their tiny, semi-aquatic, fellow river dwellers the water voles, which are easily wiped out by a spell of damp weather, floods or the sudden arrival of a predator. Being at the top of the food chain helps, but, more importantly than that, the otter frame is honed for survival. You might have thought that, for an animal constantly in and out of the water, blubber was the key to insulation, like, say, a seal, but otters carry just 3 per cent of their body weight in fat. In seals it is more like 25 per cent, which is close to that of the human body. So what is it that keeps otters warm and dry?

A clue lies in the luxuriant fur which the otter, when not

swimming or sleeping, is frequently grooming. To start with, otters generally look close to black when wet, but actually, when dry, their fur is more of a brown colour and incredibly soft to the touch. Kuschta's Californian sea otter cousins have the densest fur of any animal on the planet, at 140,000 hairs per square centimetre. On its own, that figure doesn't mean much, but when you reckon that the arctic fox, one of the most durable survivors of polar winters where the temperature may drop to as low as -20°C (-4°F), has fur with a mere 20,000 density, then the British otter, even though it trails the sea breed at 80,000, has a pelt for life. If you are wondering about us, a density of 300 on the human scalp tells you everything about the importance of hats.

Kuschta's fur, like that of a polar bear and other mammals which are constantly immersing, is dual-layered. The top layer consists of outer guard hairs, which are thick and long and form a partial barrier to keep the fur beneath dry. As she climbs out of the water, Kuschta's pelt will look anything other than smooth and sleek, but rather spiky, the coat almost like that of a hedgehog. This is because the hairs combine in little arrows to draw and drain the water off the body. But it is the underneath, the under-fur, the really dense stuff, where the otter is so truly well adapted for life in a river. Regardless of any density quotient, the ultimate test of the under-fur is that in whatever direction you stroke or pull at the half-inch hairs, you will not see the skin below.

However, the fur alone is not enough to insulate the otter. It takes one other crucial component, namely air, which, once trapped between the hairs, keeps out the cold − a sort of mammalian double-glazing − hair, air and then more hair. Grooming is, of course, about drying out and keeping clean, but it is equally for puffing up the under-fur to let air back in. And it's a lot of air. To track an otter swimming under-water, watch out for a tell-tale line of air bubbles rising to the surface – wherever the otter goes, the bubbles will follow. For otter hunters it was a tracking gift from heaven and some-thing the otter cannot avoid, for the bubbles are not created by breathing but by the pressure of the water gradually squeezing the air from the pelt. And because of this there is a definite time limit as to how long an otter can spend in water. After about half an hour the air has gone and both layers of fur are completely sodden, the cold water pressing against the skin, sucking heat from the unprotected body. Again, otter hunters knew this. If they could keep the otter immersed for long enough, regardless of how many times it surfaced for air (they can hold their breath for up to four minutes but rarely do so for more than 30 seconds), hypo-thermia would set in, hastening the end. Conversely, in the heat of a summer day, lounging on her couch, Kuschta would be content to leave her damp pelt well alone, the slowly dissipating air allowing her to keep cool, her coat fluffing and lightening to a more roan brown the longer she lay.

To find her place within this singularly convoluted riverine society, Kuschta established a pattern whilst alone during the summer and autumn months, exploring and testing each new place she came across. Sniffing tell-tale marks, she was now old enough to make a swift decision whether to move on or stay. Moving on was always her first choice because it involved least risk of confrontation, whilst staying was more nuanced, but on one particular night her luck changed.

Kuschta sat for a while at the base of the ash tree at the end of the promontory where the river split; she was perplexed. Everything she'd ever learnt told her this junction pool was a classic boundary marker, a feature in the river landscape that all otters would recognise and mark accordingly. But she had been up and down, crisscrossing the entire isthmus to check for signs of other otters, and ... nothing. So if her nose told her nothing, maybe her ears would reveal some other presence, for otters have acutely sensitive hearing. To look at their tiny ears you wouldn't think so, but I've frequently learnt this to my cost – until they dismissed it as benign, the slightest click of a door lock from my mill in the dead of the night was enough to scare the otters away before I ever saw them. No wonder that for me and so many others they can be such elusive creatures.

Otter hearing breaks down into two categories: in air good, in water bad. Despite water being such a good conductor of soundwaves, otters are not adapted for underwater hearing like, say, a dolphin or some seals, and in fact the auricle – the furry, half-moon visible part of the ear outside of the head – has a reflex response, closing over the ear hole when the otter submerges. Beneath the surface, touch, smell and sight are more potent senses. In air it is an altogether different story because otters have a range of hearing that is far more sensitive than that of humans, taking in a spectrum of high-pitched sounds something akin to that of a dog. For a creature that lives by the night and which has vision limited to movement and blurry shadows, acute hearing makes perfect sense.

After a long while of waiting, Kuschta had heard nothing beyond the usual night sounds, so with no real way of telling which was best, she took the right-hand fork; it was smaller than the other side but the stream appealed to her, the overgrown banks seemingly untrodden by human or otter. From her low viewpoint she couldn't see far, but the dense reeds, interspersed with stunted alders crowding in up to the river edge, gave her a comforting sense of protection. At its narrowest point the reeds had fallen across the stream, supporting each other at the middle, creating a cathedral-arch-like tunnel that stretched into the darkness.

The mass of vegetation barred any further progress along the bank, which suited her just fine. Otters are not ideally built for walking or running; watch them move anything faster than ambling pace and you'll see their back end rise and fall in a sort of jerky, lolloping, uncomfortable way. You worry that with all the strength in their hindquarters the front end won't stand the strain, or at the very least the otter will tumble into an involuntary front roll as the rear end overtakes the front. Otters are a bit of a mammal oddity; a creature that lives on land but is really best adapted to life in the water. They clearly know this – in all the years I've lived amongst otters they have never tried to outrun me for any distance. Actually, though I'm certainly not an Olympic athlete, they couldn't anyway, as the top otter speed is little more than fast human walking pace. When in flight, the otter tactic is clear – head for water as fast as you can. It always amuses me because they go from sheer panic on land to total confidence in the water within the blink of an eye. Once immersed, rather than swimming away into the distance at speed, they'll take a couple of strokes to where they feel safe, surfacing to gaze back at me as if to say 'Fooled you!' before disappearing off.

Unless in flight, an otter is not the type of animal that hurls itself into the water. The act of movement from land to water is one of the most fluid motions you will ever see in nature. It is a sublime rendering of evolution: that moment when a creature is so totally in harmony with the many elements it occupies that you can only marvel at the grace. To say an otter pours itself into the river is no exaggeration;

it is almost as if the water parts to welcome the creature home. Slipping into the water creates no alarm. Draws no attention. For an animal that relies on the element of surprise to hunt, stealth is no bad thing.

It was with this sinuous action that Kuschta moved from land to water, content in her own mind that, for once, she was perhaps entering vacant territory. Otters may be elusive but they are big, leaving distinct trails in the landscape. They may be able to enter water like a spoon sliding into honey, but getting out is something else altogether. Steep banks, together with that long, heavy frame, leave their mark – a few ins and outs at the same spot creates a slide, a muddy runway in the grassy bank. Kuschta saw no such slide. Otters are great creatures of habit in the routes they follow. Kuschta made the simple evaluation – no slides, no otters.

Cruising upstream with flicks of her webbed paws, Kuschta is barely visible, leaving only a slight silvery wake through the tunnel of reeds. As otters swim, at least nine-tenths of the body is below the surface; only the top of the head protrudes, with the nostrils open, eyes wide and ears pricked, ever alert. Sometimes, when extra effort is required, the rump and tail will appear above the surface, giving the impression of a three-humped creature, but for the most part the strong legs and tail stay beneath the surface, giving propulsion enough. Though perfectly at home in the river, the otter is an alien and ominous presence to most other river dwellers. They may be silent and invisible to people, but Kuschta's entry into the stream created enough distur-bance to frighten the small trout that were using the cover of night to feed away from the preying presence of larger fish. Alarmed by her arrival, they fled for cover amongst the roots of the reeds. All this Kuschta sensed through her whiskers, the tiny vibrations at first close, then far. But she paid them no heed. She was after bigger prey. For now she needed to move on, so as her own vibrations faded the trout headed back out in her wake to continue their search for food, life returning to normal.

After a while the reeded, scrubby bank gave way to open meadows, the river now wider and shallower. Shorn of any bank cover, the moon shone down directly on the surface, lighting a bright path of water ahead. Other than Kuschta, nothing moved or stirred. She was in that nether time of night when the chill had settled and the bats and owls were back in their roosts. The grass had turned cold damp, enough to send the furred creatures to ground — rabbits and voles care little for getting wet when there is scant prospect of warmth; they would be waiting for dawn before reappearing. The isolation suited her just fine, so she swam on strongly against the steady current, putting distance between her and the stream junction. As the river meandered first this way, then that, she began to tire, getting a little cold herself. Easing up onto a tree root that dipped down into the water on the inside of a sharp bend, she took stock.

Otters are rarely idle. Kuschta surveyed the river whilst all the while grooming herself, both the effort and the effect gradually bringing the warmth back into her body. Perched on the root, dried and rested, she could afford to take her time − she had chosen the warmest spot on the river. Otters exploit the smallest wrinkles in nature; her perch was one such wrinkle. With certain rivers, at particular times of the year, the water is considerably warmer than the air, and where the two meet a blanket of warm air, a layer no more than a foot or so thick, hugs the surface. If you ever see a 'smoking' river, that's the evidence, and by inserting herself beneath the mist, close to the water, Kuschta was exploiting nature's very own greenhouse effect.

Everything about the bend in the river screamed fish; the tapering flow on the inside bank on which she sat had just enough slack water where a fish might rest. The faster middle would be empty, the effort/reward ratio too much for any fish to bother to hold station in. The far bank, with its under-cuts, back eddies, tree roots and depth, was fish heaven – they could lie in there night and day, ready to dart out to any food that drifted their way. It was time for Kuschta to make a move. Slipping into the water, she let the current carry her down-stream for a few yards, then simultaneously turned and dived, pushing herself hard and fast along the inside bend, as close to the gravel river bed as she could – the closer she stayed to the bottom, the fewer options the fish had to escape; they could go left, right or up, but not down. In the dark she could see very little. But no matter; her whiskers were doing all the seeing.

She couldn't be sure, but maybe a fish had darted off into the distance. She let it go. Her hopes were really pinned on the far bank. By her estimation the fish would be facing upstream, looking out for food, so if she came at them head-on she'd have a crucial moment of advantage as they had to turn to flee. So surfacing well upstream, she coughed in that way that otters do, noisily sucked in air and dived. Her swimming, combined with the current, moved her fast. For a short moment the fish didn't notice her coming, her movements masked by the midstream current. But then in total alarm they knew there was something dangerous amongst them. Kuschta's whiskers were alive with informa-tion. She lunged as a trout slipped for cover under a tree root. Her teeth grazed another that bounced off her head. She accelerated after a third but it had too much of a head start. Undeterred, she surfaced, paddled for a moment to recover and then headed down deep, preparing to get amongst the roots this time.

It is something of an irony that any fish would be best advised to do nothing when hunted by an otter; by remaining still, the chances of detection in the dark, swirling water would be next to zero. But flight is the natural instinct of fish, so otters exploit this primordial reaction to danger. The fact was, Kuschta had no exact idea where the fish lay, she just knew that they were probably hiding in the cavity beneath the tree roots, and if she could spook them their instinct for flight would lead her to them.

Rising from below and slipping between the trailing roots that hung down in the water, Kuschta's bulk filled the confined space. Any speed advantage the fish might have had over her was gone as she drove them down a watery cul-de-sac. Was it two, three or four fish? She couldn't really tell, such was the confusion as they tried to push past her to the safety of open water. However many it was, it didn't really matter – she needed just one, the currents of vibration honing her in on a fish trapped between her and the bank. The soft belly of the fish gave a little as she made contact with it with her mouth before she drove her long, curved canine teeth into the flesh. The now-wounded trout flexed head and tail in unison to escape the pain and capture. Reversing out, Kuschta kept her jaws clamped tight, the backwardly curved teeth maintaining a certain grip on the struggling fish. Breaking the surface, Kuschta's nostrils flared open to breathe in air, whilst the trout splashed and crashed about her head, in a flailing death throe now exposed to the same air. Swimming back across the river,

Kuschta headed for the root perch, scrabbling up and out, sending a spray of water all around as she delivered the *coup de grace*, violently shaking the fish to snap its spine.

Kuschta didn't bother to groom or preen; she ate as if her life depended on it. By the time the first fingers of the cold winter dawn showed across the meadows she was done, the leftovers just a ragged tail. It was time to hide. Her eye was drawn to a mess of dried reeds and twigs that had been gathered up then left behind by a recent flood, piled up against the base of the tree. Pawing at the pile, she exposed a gap in the web of roots at the base of the tree. Squeezing through, she found a small cavern beyond, the sides and roof made up of old, gnarled brown alder roots, most of their growing done. The floor was softer, still alive, a bed of little pink nodules ready to sprout in the spring. Dragging some of the leaf litter inside, she circled around as best she could in the tiny space, fashioning a comfortable mattress which she nestled into. Sleep was not long coming, but before Kuschta finally drifted off she sensed she might finally have found a place to call home.

CHAPTER 3

SOMETHING IN THE AIR

Winter

I've lived on and around rivers pretty well all my life, but it wasn't until my fourth decade that I finally saw an otter. And even after all that waiting, that first sighting wasn't under particularly auspicious circumstances.

I had just bought an abandoned water mill that straddles a small chalkstream in southern England, called Wallop Brook. It did, and still does, comprise two buildings – the miller's cottage and the mill building. The former was just about habitable and the latter was really nothing more than a foursquare brick structure rising over three storeys, completely empty bar one important element: the mill wheel itself. I gleaned from the villagers (not all overly friendly when I first moved in ...) that the corn-grinding mechanism had been stripped out years before, the last production sometime soon after the Second World War. A few things remained to remind a casual visitor of a past that stretched back over a thousand years – you will find the Nether Wallop Mill listed in the Domesday Book. The side wall of the building was hung with slates, faded white signwriting emblazoning in two-foot-tall letters the legend F. VINCENT'S NOTED GAME FOODS. The mill had produced both bird food for a wider market and, on a lesser scale, flour for Nether Wallop and the surrounding villages. Out in what is now the garden, where in the past sheep grazed up to the back door, there can be found a complicated array of a mill pond, pools, hatches, carriers and relief streams. It might look antediluvian to us today, but in Mr Vincent's time, and long before that, too, these old-fashioned devices controlled the flows that were vital for driving the water wheel and sustaining the milling industry. In more modern times, and for my purposes, they are far from defunct, their control being the difference between me having a wet or dry house in times of flood.

As I write this today my feet are poked under a giant cast-iron spindle, the central core to an even more giant cast-iron mill wheel, the height of two men, that is separated from me by a low wall, topped by a glass partition to the ceiling. Effectively my office is divided in two – one half for me and the other half for the mill wheel. Despite the constant pummelling roar of the water next door (yes, every minute, of every hour, of every day, year in year out), I chose to build a desk over what used to be the drive mechanism for the grinding stones. If I look up I can see the marks in the ceiling beams where the power take off gear connected to the spindle to the grinding gear. Behind me is an old-fashioned winding handle that turns two cogs, which in turn lift an iron gate that controls the flow of water into the stream that powers the mill wheel. I only need turn the cogs two or three notches and the wheel will turn. It is a slow, powerful, creaking turn, the thirty-two buckets (the official term of a mill wheel paddle) taking nearly a minute to go full circle. I have to remember to keep my feet clear of the turning spindle when it is in motion.

But it wasn't always like this. When I first arrived, the wheel was stopped and had been that way for years. In some distant past it had slipped out of level alignment; for a while it obviously continued to turn despite being out of kilter, cutting circular gouges in the wall that are still plain to see. But at some point it must have jumped out of the shoe in which the nub of the spindle sat, to lean at a crazy angle jammed up against the wall. The iron control gate had rusted away to nothing, the cogs that raised and lowered it long gone. You'd think that it was a hopeless case. Plenty suggested that I might just as well sell the iron for scrap. However, I am not that easily deterred. Believe it or not, there are still skilled wheelwrights working today. Men with boiler suits, toolboxes full of mighty spanners and hands perpetually ingrained with grease. They took one look at it, pronounced it sound and returned some weeks later with newly made parts that made the wheel operation whole again.

You might wonder why this is relevant to my first otter sighting. Well, there is a vaulted tunnel where the mill straddles the river, carrying away the water after it has powered through the wheel. After years of disuse that tunnel was virtually blocked. We had donned waders to check it out, jammed as it was with logs, mud, brushwood and all sorts, but really it was too dark and confined to tell much. I was all for some extreme raking to clear it, but the millwright guys assured me that the water would do all the work. So with great ceremony the iron gate was lifted for the first time in decades. The water flowed, the mill wheel turned and the tunnel gradually filled with water until the force was so great that a plug of ancient detritus burst through into the mill pool below. Suddenly the pool went from shallow and clear to deep and dirty. Tree roots, bald tennis balls, reeds, twigs and all sorts swirled in the surface, but something in the back eddy caught my eye. It looked like an over-inflated, half-sunk, part-hairy, grey and pink balloon. I dragged it closer with a stick.

I guessed it was something dead. At first I assumed it was a badger, but the long tail, denuded of hair in death, told another story. As the corpse flipped over, it was clearly an otter. I am no pathologist, but years of living in the country usually gives you some ability to tell what a creature has died of, or been killed by, but this otter was too far gone for any postmortem. The fur was peeling away, exposing the greying pink skin beneath. Bones were showing through the flesh of the legs. I suspect in a week or two it would have been unrecognisable even as an otter. So I can only surmise as to how it had died. In all probability it had crawled into the tunnel as a last place of refuge, hit perhaps by a car, which is common enough. Or maybe it was on the wrong end of a fight. Or perhaps it was simply old age. Whatever the reason, it was a sad way to see my first otter.

I must admit, at the time I didn't think very much more about it, putting it down to a freak occurrence, but as I spent more time at my desk beside my newly refurbished mill wheel I started to have unexpected company. As I mentioned earlier, the wheel housing is a separate room of the mill, through which the river flows, splitting into two channels. One channel takes up about two-thirds of the width, over which hangs the wheel itself. The other third is the mill race, where the water pummels through. The race is a sort of relief channel through which the river is diverted when the wheel is not running. The whole wheel housing is effectively open to the elements with brick arches over the river at either end of the building. On the upriver, or inflow, side, two huge, ancient oak beams straddle the width of the room from which are hung the iron gates that control the flow through the two channels. It was these beams that the otters adopted as couches.

I say 'the otters', but I really have no true idea whether it was the same otter who arrived often or a series of visitors. The sightings were nearly always fleeting as I came into the room to sit at my desk; a blur caught in the corner of my eye followed by a splash. At first I ignored it, thinking it was, well, I don't really know what I thought it was. A mink perhaps, or a stoat; they are far more common. Even a rat maybe. But one day when I was adjusting the control gates I saw shining atop one of the oak beams what today I would instantly recognise as a spraint. Back then, less so, or, if I'm being honest, not at all. A trip to my desk and a visit to Google put me right. I determined to be more discreet when entering the office next time.

However, my definition of discreet and that of an otter is a very long way apart. Two or three steps into the room was only ever the best I could do before the splash and the rapid departure. I did take to rushing outside to at least have the satisfaction of following the bubble trail as the otter headed off underwater. Sometimes he, or it could have been she, would surface to look back, but generally the last I would see was a wet sliver of fur slide itself over the weir and disappear into the pool below.

A few times I did get closer. One summer afternoon I went

into the mill wheel room, blinking as I went from the bright sunshine to semi-darkness, only to be struck rigid at the sight of two otters sitting on the oak beams. Who was more shocked I have no idea. I looked at them, they looked at me. I didn't move but they did, twisting and diving into the water, fleeing at speed. The other times were when I worked very early or very late at my desk. I'd hear some splashing and coughs of exertion as an otter hauled itself out of the water using the ironwork as a sort of ladder to perch on one of the beams, grooming and generally making itself comfortable before settling down. It was then, and still is now, a wonderful thing to see up close. Occasionally the otter would spy me, our eyes meeting and the reaction variable. Sometimes instant flight, other times mild curiosity before choosing to ignore me. The latter was fine by me. Working with an otter peering over your shoulder is an oddity worth getting used to.

It might seem odd that an otter would choose the mill wheel as such a regular stopping-off point, being, as it is, in the midst of a human habitation. But I think it is something of a combination of things that makes it so attractive – the antiquity, the lie of the land around the mill, the location and, more recently, an awful lot of fish. There is no doubt that they have been using that oak beam as a couch for a very long time. Spraints are not just odorous but are also pretty toxic in dung terms. Regular sprainting spots on grassland will turn the turf brown then dead. It will really take the ground a long while to recover, the deposits having much the same effect as spilling fuel or oil on your lawn; once you know what to look for, it is an easy way to tell whether otters are around. In a similar fashion, otters who live by the sea will take a particular liking to a prominent rock or outcrop. Clearly the spraints can't do much damage to solid stone, but the spot will turn green in time, much like the copper roof of a church. Back closer to home, my oak beams have suffered a slightly different fate; each now has rotten indentations where the otters have laid down their marks over the years.

The land around the mill is a regular Spaghetti Junction of water courses; not only does the water go under the building but it goes around it on both sides – we are effectively moated. To put that into some sort of perspective, imagine you are looking directly at a rugby ball; from the top the three lines of stitching represent where the single river is split into three. Down the left goes the original Wallop Brook, a fast clear stream that burbles over gravel. Down the middle is a much wider, deeper slower river which we (confusingly) call the Mill Pond. It is this that drives the mill wheel, which is where the rugby ball laces would be. Down the right is a side stream, or carrier, a man-made channel that was created to regulate the level of the Mill Pond. All have been dug or adapted by man in past centuries to manage the water flow, with the addition of some connecting channels that run crossways between them. Downstream of the mill, at the base of the rugby ball, if you like, all three come back together where a united brook continues on its way into the water meadows.

All in all, this is otter heaven; when on land, there is no point at which an otter is ever more than a few bounds from the safety of water, and they do treat the respective streams as regular highways. I can see from the permanent tracks in the grass and the slides that they arrive via one stream, cross by land to another, tracking back to the original one further downstream by a different route. They barely deviate in the routes they follow; in the spring the fresh grass is pressed down, by summer it is pounded brown and in winter there is muddy track. And then, of course, there is the snow. They are, if nothing else, creatures of habit.

The mill is also on the edge of two of the Wallop villages that stand along the brook, our building being the first or last outpost, depending on your direction of travel. The two settlements, Over Wallop and Nether Wallop, like the territory of otters, are very linear. The ancient meaning of the word 'wallop' is hidden valley, and the combined villages stretch about three miles, the homes of just a few hundred people mostly hunched up close to the course of the river. I suspect that the mill wheel, the last stop after all those miles of habitation, is where otters can arrive and depart by water, almost like a proper holt, which must seem like a blessed refuge. Conversely, if they arrive from the direction of The Badlands (more about this place in a moment), after a trek over four or five miles of wild and barely habited river, the stopover with us must appeal for different, but equally important, reasons.

The one thing I haven't mentioned is the trout lake, which for all the obvious reasons makes us an undoubted attraction on the itinerary of any otter. The lake, which lies just 35 yards to the west of the mill (to the left of that imaginary rugby ball) is fed by offshoots of the Wallop Brook that flow in at the top and out at the bottom. It is the shape of a kidney, which size-wise would more or less fit into a football field. There are grilles at the inflow and outflow to stop the trout escaping, but it is otherwise unprotected, just part of the landscape. But this is not really your normal lake. It is stuffed full of rainbow trout, because this is where I teach fly fishing – with new people coming every day you need a heavy density of fish, and during the season, April to October, the stock is replenished fortnightly from a local trout hatchery.

I don't like to diminish the status of the rainbows; they are hard-fighting fish that are great to catch and in their native North America they are wily survivors, but here, when the fishermen have gone home and the night falls, the odds are stacked against them when the otters come calling.

During the spring and summer when I go out to do my early morning rounds, clearing the sluices and adjusting the hatches in preparation for the fishing day ahead, I expect to find a fish corpse, or the evidence of one being caught, more or less every other day. Usually it is a victim of an otter, though occasionally it is a heron, but it is pretty easy to tell the difference. If it is an almost whole fish, the heron will have left tell-tale stab wounds. Conversely, if the bird has had time to eat pretty much all the fish, it will look more like a cartoon fish skeleton, the left-behind bones picked clean. Otters, on the other hand, generally start from the head down, eating everything, bones and all, as they go. In the depth of winter, when food is scarce, it is unusual to find part-eaten trout – protein is too scarce. It is only really in the summer, or when the mother is teaching the pups to fish, that otters abandon a trout without finishing it off. Sometimes I have to look really hard to see whether they have been, the only evidence those few flecks of blood or bright scales similar to what I saw on that snowy morning. I suspect the otters had been robbing me blind of trout for years without me ever knowing it.

As winter clutched at the throat of the countryside, daily squeezing every last drop of life from the less hardy inhabitants, Kuschta took to exploring her new territory. Unfettered by the constraints of other otters, she was free to move at will, marking the land along the Wallop Brook from the junction pool to the headwaters, where it is barely a river at all as the bright crystal water springs from the ground. If you flew up the valley like a bird you'd see that, despite all the apparent habitation – houses, farms, roads and all the other things that civilisation brings in its wake – the Wallop Valley is surprisingly wild. Woodland crowds up to the bank for at least a quarter of its length, hiding the river from prying eyes. Water meadows, rough-grazed by cattle and flecked with wild flowers, merge the land with the water. In some places it is just a river lost in a wetland swamp. We call this lost place downstream of the Mill The Badlands, where reed beds, criss-crossing rivulets, soft soggy ground and a scruffy, fallen willow plantation look like a terrific mess. It is rarely visited by people. Sure, there are some tidy gardens that come up to the edge of the brook in places, bits that have been adapted for things like my mill or banks that have been realigned to prevent flooding, but on the whole it is a natural stream that hasn't changed much in the past two or three centuries.

When we think about the history of our landscape, it is strange that otters don't feature more in British folklore, history and culture, for they have been part of our lives since the first moment man made settlements on the banks of a river. From that time onwards, as we invaded the territory that they had called their own for millions of years, otters were amongst us but never really part of us – mysterious creatures that we saw rarely and understood even less. The inns along the highways of Britain are testament to this absence; the names The White Hart, The Black Horse, The Bear, The Swan, The Bull and even The Black Rat offer an insight to the creatures that have impinged on our culture down the centuries. But The Otter Inn? Well, there are some, but very few considering it is our largest semi-aquatic mammal.

The more you think about it, the stranger it is. After all, otters are not exactly small; nose to tail they are close to four feet long. A fully grown male weighs around twenty-two pounds – that is heavier than a terrier or about the same as a beagle. In feline terms, think twice the weight of a healthy cat and twice the body length. And a river through a town is a much-watched place – you'd think they would hardly go unnoticed, plus you'd expect that the numerous opportunities for food would draw them into human orbit. Rats and foxes have adapted to human habitation, thriving on our detritus and finding homes that man has, by accident rather than design, created for them. But not otters. They seem to shun the opportunities afforded by man, even changing their habits to become yet more secretive.

We think of otters as nocturnal, but they can equally be diurnal – active by day instead of night. On the south and west coast of Ireland otters regularly swim past anglers during the day; visitors are astonished, whilst for the locals it is so common as to pass unremarked. It is the same in the Scottish Isles, suggesting that where people are sparse otters are content to alter their behaviour accordingly. When they choose the night, they do it to avoid their greatest adversary – man.

Maybe there was a time long, long ago when man and otter lived in perfect harmony. After all, nobody ever seems to suggest that otters make good eating. They were not hunted for food, unlike the slow-witted beaver who, also native and incredibly populous to Britain at one time, was hunted to extinction as soon as early man took to living in the river valleys. In fact, the only people who seemed regularly to eat European otters was a group of Carthusian monks in Dijon, France, who stretched the truth to get around some awkward theological dietary requirements. Banned by holy order from consuming meat, they cunningly deemed the otter to be a fish. Now whether this was because it ate fish or lived like a fish, nobody is exactly sure, but accounts of the time rated the flesh 'rank and fishy', so the monks must have been somewhat desperate.

So aside from a few monks, maybe there was a time when the otter went about its daily life without a care in the world. A time when the fish were plentiful and the people few, when otters were free to range over huge tracts of unsettled land where the rivers were wild and the woodland dense. A time when otters feared nobody and wanted for nothing. It is a lovely thought; a sort of aquatic Garden of Eden. But if such a time ever existed it most certainly came to an end in the Middle Ages, when the population of Europe increased. Communities coalesced around rivers, the fertile valleys were gradually cleared and drained for agriculture. What was done a thousand or fifteen hundred years ago across southern England was not so very different to what is being done to the rainforests of South America today. The destruction of a habitat that slowly marginalises the indigenous species. Some will survive this change, others will become extinct. A few will become mortal enemies of man; unwelcome at best, feared at worst. The history of medieval times tells us that the otter fell into the 'unwelcome' category, labelled as the 'fish-killer', stealing food from the rivers that 'rightfully' belonged to the more 'deserving' mankind. It is a tag that remains today, but the persecution dates back many centuries.

The more you look back, the more astonishing it is that otters have avoided extinction in the British Isles. We might think of the eradication of a species as a rather modern manifestation of human behaviour, but otters have been on the hit list for over a thousand years. Way back in the twelfth century society went to war with the otters and lutracide was born. Henry II appointed the wonderfully titled King's Otterer, who was charged with the extermination of the species. It was no passing fad; this was serious business. With the title came a manor house, land and an annual stipend all bundled up in legislation to create the Otterer's Fee. The first Otterer, a man called Roger Follo, from his 'Fee' in Aylesbury, Buckinghamshire, went about his task with a new form of otter control, namely an otter-hound pack.

However innovative and hard-working the Honourable Follo might have been, any success must have been transitory, for by the fifteenth century Henry VI was back at it again with the creation of the Valet of our Otter-Hounds. But otters continued on their merry way until 1566, when, frustrated by their continued existence, Parliament passed the Acte for the Preservation of Grayne, which classified otters, along with badgers, foxes, hedgehogs* and others, as vermin, allowing parish councils to offer bounties for their capture. Sixpence, the reward for a dead otter in the early 1600s, strikes me as a lot of money and gives some indication of how otters had become a significant public enemy.

It is interesting to ask why otters were elevated to this status. I think we can say with some degree of certainty that their fate as public mammal enemy number one was cast for the next three centuries in 1653 when Izaak Walton wrote about them in *The Compleat Angler* – a huge bestseller when it was first published and subsequently one of the most reprinted books of all time. He declared,

'I am, Sir, a Brother of the Angle, and therefore an enemy of the Otter; for I hate them perfectly, because they love fish so well.'

---

* You might well be re-reading the list and wondering why on earth hedgehogs were included. Apparently they used to steal milk by suckling on the teats of cows, goats and sheep. I think this probably tells of legislation borne out of prejudice rather than fact.

This is pretty stern stuff for an animal that carried no disease, kept clear of people and posed no physical danger. But the fact is that otters were eating the fish owned by those who held the reins of power: the monarchy, noblemen, the church and the educated. These were singularly bad groups to antagonise. Noblemen owned the rights to fish rivers, which was an important source of income and food. Fishing grounds were jealously guarded – not just physically but in law, for they were specifically mentioned in the Magna Carta. The draconian law that went as far as capital punishment was enough to keep the commoners at bay, but otters required something else. Monasteries and the palaces of bishops had for centuries reared fish in ponds, but they were difficult to protect and made tempting pickings for a hungry otter in the depths of winter. Then people such as Walton discovered the joys of angling as a pastime, which pretty well sealed the public perception of the otter. Whether they truly posed a threat to fish stocks is debatable, but the fact remained that otters had got on the wrong side of the wrong people.

So the notion of the otter as a quarry became entrenched in the psyche of the nation; along with foxes and deer, the hunting of these animals with hounds was an accepted pastime. It was both part of the social fabric of the British Isles and a requirement for the management of the country-side, albeit the latter of dubious value. You'd have thought that as feudalism gave way to industrialisation society would lose interest in the otter, but not a bit of it. In the Victorian era, otter hunting became quite the fashionable pursuit, reaching its zenith in the years between the two World Wars. However, for all its barbarism, twentieth-century hunting barely put a dent in the otter population. Ironically, it was the hunts, with fewer otters to hunt, who first alerted a wider public to the decline in their numbers across post-war Britain, as over two decades – the 1950s and 60s – otters all but vanished from the countryside. Hovering on the brink of extinction, the search was on for the otters' insidious foe before it was too late.

What *has* changed over the past half century in our country is the otter population. Wind back the clock eighty years ago or more and it is a fair bet that Kuschta would have faced fierce competition along the Wallop Brook, with probably just two or three miles to call her own compared to the nine miles over which she ranges today. The truth is that otters are just clawing their way back from the edge of extinction.

It really was a mighty achievement of twentieth-century man to bring otters to this sorry point in time, where their very existence was threatened. After all, we have succeeded where centuries of persecution have failed, but we did it entirely by accident, and then in recognising the ongoing damage we failed over successive decades to put it right. It will be of no comfort to know that we were not alone in this. Across Europe – in Germany, Italy, the Netherlands, France, Finland, Sweden – and in fact in just about every mainland country, we have seen a catastrophic decline in the population of otters in the post-war era. A culture of persecution continued to play a part; in Switzerland there were 40–60 otters left when given protection in 1952. By 1960 they were all gone. To give you some idea of the level of hatred, three captive otters in Zurich Zoo were killed by visitors. But ultimately it was a poison, spread in the name of progress, that took otters to the brink.

Seven decades on from the end of the Second World War, it is hard fully to understand the mind-set of a Britain traumatised by a conflict that had kept the nation on the brink of imminent starvation. What we would now call food security, the ability to feed the population with crops grown on home soil, was the mantra of all governments of all hues in the years immediately after the war right through to the 1970s. As the Minister for Agriculture, you would have been one of the top five men in the cabinet; today you would be an also ran. The National Farmers Union held sway at every level of decision making in the drive to boost food production. The BBC joined in, *The Archers* a handy propaganda tool for agricultural lobbying. What was good for farming was good for the nation. Where nature stood in the way of progress, science was enlisted, the upsides lauded and the downsides ignored. Intensive agriculture, the please-all, cure-all of the time, required chemical intervention, and so it arrived in 1955.

It was the simplest of desires that caused the first problems; the wish to protect newly sown corn from pests for better germination rates. Coated with an organochlorine pesticide, the effects were almost instant − wheat and barley thrived, bringing marginal arable land into production and boosting yields. The trouble is, fields don't exist in a vacuum. Wood pigeons and songbirds eagerly scratch out the newly planted seed from the ground, consuming it in quantity. They were the first to die, killed by direct ingestion. Next up were the species that died from eating the dead. Foxes and barn owls were hit hard, but it was the dramatic decline of the peregrine population that sounded alarm bells in 1956. The Royal Society for the Protection of Birds (RSPB) started to investigate the avian deaths, fingers were pointed but there was unstoppable momentum behind the use of organochlorines.

They were used in a multiplicity of ways that spread them into nature's food chain: sheep dips, bulb dressing, orchard sprays, timber preservative, moth-proofing fabrics and carpet-making, to pick a few. It might seem a long step from those processes to killing a top predator like an otter, but when you consider, for instance, that great rug weavers like Wilton built their factories by rivers for water and for waste disposal in an era when environmental legislation was all but non-existent, then the connection comes into focus. So, as the invisible fingers of pollution touched just about every river (sheep dips were particularly pernicious in this respect), the problem turned into the unknown crisis, with nobody really noticing through a combination of bad luck, the secret nature of the otter life, the delayed effect of the poison and inaction. The bad luck came in the form of a report published in 1957 but based on data from 1952. Why there was a five-year delay I have no idea (though conspiracy theorists might), but it concluded that the otter population was doing fine. With the birds to worry about, nobody gave much more thought to otters and, being such secret animals, few had any real idea what was happening to the population as the insidious chemicals did their worst. This is how it happened.

Being top of the food chain is all very well, but the implication is that you prey upon everything below you. That's fine just so long as your favourite foods – eels and fish, in the case of otters – are good to eat. By the early 1960s this was far from being the case. Eels, which live for 10–20 years in ponds, were absorbing the organochlorines into their bodies at a rate of knots from a diet of similarly affected grubs, earthworms and insects. The same thing was happening with fish from their diet of invertebrates: nymphs, snails and all those other bugs you find in a river. But the chemical pass-the-parcel wasn't killing outright the otters or the things they ate. There were no corpses littering the river bank – if there were, things might have turned out differently. No, otters were hit hard because, with little body fat to act as a buffer like, say, in the eels, the sub-lethal poisoning went straight to the reproduction organs, slowly rendering the population infertile. Otters were not dying, they were dying out.

It is a hard case to make from an emotional standpoint, but it was otter hunts that were the greatest guardians of *Lutra lutra* during this time. They had a vested interest, that was true, but nobody was closer to the lives of the otter. It is counter-intuitive, I know, but when it came to habitat protection and preventing uncontrolled extermination, the hunts were the otters' best friend. One hunt in Dumfriesshire even went to the lengths of importing otters from Norway for reintroduction into the wild after a localised population crash. By the early 1960s the declining population was more than just a local occurrence; packs up and down the country were reporting fewer and fewer otters. Some packs closed down. Others hunted mink instead. The remainder changed their method of hunting, reducing the kills from 50 per cent of all otters chased to 15 per cent, limiting, as far as it was possible, those kills to old or sick otters. By the time otter hunting was finally banned as part of an Act of Parliament that gave the animals protected status in 1978, the fifteen otter packs that remained were killing just 150 otters a year between them.

The threat of extinction was never just from hunting, but as the news of the otter decline filtered through to the wider population during the 60s and early 70s this was what took the brunt of the blame, as the anti-hunting lobby gained a voice. Other voices called for investigations, and reports were duly produced. Water quality, habitat destruction, disturbance through human activity and even the lowly mink were the four reasons generally cited for the decline of otters, but rarely was the systemic poisoning given the prominence we now know it deserved.

However much it was wrong, it was hardly surprising that mink took part of the rap; a non-native species first imported in the 1920s, it had adapted to life in Britain pretty well, the population gradually expanding over time, with regular boosts from escapees from mink farms. The European mink, *Mustela lutreola*, are, as the second half of their Latin name suggests, related to otters, part of the mustelid family. They are more gregarious than their larger cousin (they are about one-third of the size), and you are far more likely to see a mink than an otter as they are less wary of people, preferring to be out and about during the day. The mink were blamed because nature abhors a vacuum. As the otters disappeared, the mink expanded into the vacant space, people assuming that the mink, with a reputation for being vicious, had driven out the otters. Nothing could be further from the truth. Today, with otters in the ascendant, mink are finding themselves marginalised, and their population is declining.

Habitat destruction, mostly in the relentless process of urbanisation, will always be an issue for otters. Actually the worst of the damage was probably done in the 1940s and 50s when, again in the name of food production, vast swathes of otter-friendly wetlands were drained and thousands of miles of rivers straightened and dredged. Disturbance? Well, that was cited in the form of more leisure uses for rivers – boating, fishing, canoeing and so on – but otters are pretty tolerant of minor human incursions into their territory and no amount of daytime splashing would have had a significant effect. Water quality (aside from the organochlorines) was actually by this time going in the opposite direction, improving rather than worsening. The River Thames is often cited, reaching its polluted nadir in 1957 when classified as a 'dead' river, incapable of sustaining a fish population. Since then, along with most other rivers, the situation has improved, with salmon now regularly running up the capital's river. Confusing? Well, only if you were directed, as most people were, to look in the wrong places. But for those close to the science, otter post-mortem data was tightening the noose around the neck of organochlorines – the problem was that nobody in power was prepared to pull the lever that consigned them to death instead of the otters.

Finally, a report in 1968 that charted the catastrophic collapse in otter numbers captured the headlines, leading to a general acceptance, albeit grudgingly in certain circles, that organochlorines were the problem. However, vested interest and inaction delayed the widespread banning of their use until 1975. This you might think was a cause for dancing in the street, but they were simply replaced by the equally bad organophosphates the following year with, almost unbelievably, the original chemical continuing in use for commercial bulb farming in Cornwall and the compulsory practice of sheep dipping right through to 1992. In that same year, the authorities finally called time on organophosphates, replacing them with synthetic pyrethroids. Relief? Well, not really. The synthetic substitute, rather than infecting the food chain, went a step further by wiping out entire groups of invertebrates – so the very insects that fed the fish that fed the otters were disappearing. This new menace was finally banned in 2006.

I wouldn't be at all surprised if your head is spinning from all these dates and scientific terms, but I chart it because it is truly amazing that, despite fifty years of sustained attrition, albeit unintentional, otters are still with us today. As with everything to do with these secretive creatures, it is hard to pinpoint the exact moment when the population reached its lowest point, but most observers seem to agree that it was some time in the 1990s – by then it was estimated that otters were only present in a handful of English counties. In Wales, Scotland and Ireland, with less intensive agriculture, the numbers had held up better. But the long road to recovery, which continues to this day, had begun. It was never going to be a fast journey; the 'organos', with their various suffixes, have to dissipate gradually from the food chain. Otters are not the most prolific breeders at the best of times, their progress tied to the health of the rivers and the availability of food. Fortunately they hung on in enough places to keep a breeding population alive; the areas mostly away from agriculture and industry. And that otter society requirement for the juveniles to travel great distances to find new, unoccupied territories had started to disperse a new population nationwide.

As luck (they finally got some) would have it, they had some breaks along the way: more legal protection, better environmental oversight of the watercourses and an explosion in the crayfish population – one of their favourite and most nutritious foods. All of this culminated in a survey published during 2011 that had found otters in all the forty-eight English counties, Kent being the last piece in the jigsaw. But don't be under any illusions that the danger is over; they remain rare and under threat.

Kuschta knew nothing of her rarity, nor the perilous past her recent ancestors had trod to bring her to this point. All she knew was that The Badlands should be her home. A place good for otters, bad for people, as you'd struggle to walk across this landscape without considerable difficulty and deviations. A very long time ago this was water meadows, low-lying land in the flood plain that was deliberately 'drowned', covered by water diverted from the river during the winter and spring, to boost the growth of grazing pasture for sheep and hay making. It didn't happen by accident; seventeenth-century Dutch engineers had been engaged by the large landowners, church and gentry to dig side streams or, as they are correctly known, carriers that were regulated by wooden hatches all along the valley that enabled and controlled the flooding. If you think of a human skeleton with the river as the spine and the ribs as the man-made channels, then you will get some idea of the layout. Long abandoned to nature, the defunct Dutch engineering now defines this landscape.

So where you would struggle, Kuschta revels. To start with, you'd have a hard time even entering her domain from most points of the compass, protected as it is by thick swathes of dense, prickly hawthorn, the vicious barbed sloe bushes and bramble briars. Of course, for her this presents no problem, slaloming between the stems, free to come and go at will and unnoticed. There is one gate to The Badlands, which is not much used and is your best entrance. From the gate it is hard to tell much about the landscape beyond, as almost all you can see is reeds. They stretch ahead of you, to the left and the right, almost to the height of a man, obliterating your view of the horizon. In the winter the reeds are desiccated, bleached to a dirty cream with the grass-like seed pods furling out of the top. Occasionally a small songbird, a wren or robin, will alight to the top of the stem, swaying perilously from side to side as it pecks for scarce food.

Too wet for people or cattle, the occasional visitors to The

Badlands at this time of year are dogs; in time Kuschta learnt
to recognise the sound of their imminent arrival as the
beaters sweep the surrounding fields on shoot days. The guns
sound in the distance. The horn wails to signal the start and
end of each drive. The click-click-click and clack-clack-clack
of the beaters tapping their sticks against the trees is interspersed by shouts as they clear the woods of birds. The wild
fluttering of wings, as a pheasant takes flight, is followed by
cries of 'forward' to alert the shooters of incoming sport.
Sometimes Kuschta will feel the ground quiver as a shot bird
thumps into the ground somewhere close by. Occasionally
she'll see the last twitches of life play out in front of her.
Soon the gun dogs appear, splashing through the water and
crashing through the reeds, directed by whistles and calls
from afar. There is no subtlety to their arrival, playing havoc
with the snipe and curlew who rise fast to the air in protest,
the unwise making flight over the line of guns. In truth, dogs
of the retrieving kind are not much worry to Kuschta. At first
she had fled in terror, bolting from her couch to the safety
of the river. But now, a few months on, she knows to hold
her ground. Mostly the dogs are too intent on finding the
birds to even notice her. If one does, she'll snarl and hiss at
the barking canine until the distant handler, frustrated by a
dog going feral, calls it back with irritated shouts and shrill
whistles. Soon the dogs retreat, the noises fade and The
Badlands is at peace again.

The side streams, once the open channels that carried water
across the meadows, are now both Kuschta's paths and her
larders. If The Badlands were shorn of vegetation you could
very easily spot what are today effectively a series of parallel
ditches about 50 yards apart, radiating at right angles away
from the river, the ground between each rising and falling to

create a gentle mound. From a distance it is a landscape that might look like a soft swell rolling in to a shore. If you choose to traverse The Badlands you'd be well advised to follow along the length of the mounds – they are relatively dry and firm underfoot. If you make your path by crossing the old streams in turn, prepare for a long and tiring effort. Though not exactly quicksand or bottomless pits, these are cloying obstacles, too wide to jump and with no firm crust to support even the lightest person. But otters? Well, that was altogether a different story.

Aside from the sheep set on The Badlands for a month of grazing in the late summer, rarely did anything of any size interrupt Kuschta's rule of this stretch. Too wet for rabbits, badgers or foxes and away from human intervention, the largest, wildest thing that came through were startled deer on the run. With springing leaps they would clear the tops of the reeds, but the success of each leap would entirely depend on the landing – hit a mound and they sprang on at speed. Hit an old stream and it was all legs, mud and rasping gasps before hauling up and on the way again. It was an almost daily occurrence and Kuschta would listen out for it; as we'll discover deer were her unwitting eel finders – an essential staple in the diet of an otter.

But if the eels, however accidentally uncovered, were the regular gift of nature, the manna from heaven was provided by me in the form of the trout lake at The Mill. No emaciated trout here, every one of them nourished daily with pellets, a true living larder. A popular winter venue, I'm always surprised Kuschta doesn't visit it more often during the worst of times. It is only a mile or two from The Badlands so, as far as I can tell, far from being on the extreme edge of her territory. She could pretty well visit every night if she chose to, but despite the abundance of easy prey, she doesn't. I often wonder why. It is said otters have a higher level of common sense than most other animals when it comes to their food sources, hunting in such a way that allows the population to regenerate rather than wiping it out. It sort of comes back to one of the multiple purposes of spraining, an otter quality mark telling of what was caught where and when, providing a sort of self-regulation. But if I shared with Kuschta the truth – that I restock the lake to replace all the trout she takes – maybe she would alter her behaviour.

As it is, I'd estimate the visits are a bit more than every other day at this time of year. In my early years at the mill these losses used to annoy me; the fish corpses along the bank both a financial and emotional loss. I like fish as much as I do otters, but then I read a simple quote by I can't remember who: 'otters are rare, fish are common'. For some reason it struck a chord, and since then, though Kuschta's pillage doesn't exactly fill my heart with joy, I've accepted it as a price worth paying for the sight of a beautiful creature. That said, it doesn't stop me sending a few choice curses her way when I see a fish with half a tail or a lacerated side.

In the winter she arrives at the lake early, and that always takes me by surprise. It shouldn't really. My 'early' is judged against the human definition of nocturnal, let's say 9pm to 6am, whereas hers runs from sunset – 4pm at the winter solstice – to sunrise at 8am. You'd think that those sixteen hours would be a massive opportunity for her to travel far, but otters are not active through every hour of darkness. They tend to pack the most into the first two or three hours, resting for the remainder until the last hour before dawn when they get busy again. It is not what you'd expect, but it is sometimes easier to find otters in winter than summer; it certainly does not involve all-night vigils. I'll see Kuschta dive for cover as I nip out to catch the last post, or I'll hear that tell-tale splash as I flick on the office lights to combat the gloom. If I'm out before sunrise for my morning checks of the river hatches we'll often come face to face. The best is when she has a fish to eat. While she is intent on that, I will generally see her (the dreadful noise of her tearing apart the fish is always the giveaway) before she sees me. I can't help myself, but I always creep up as close as I can. I guess I should circle around and leave her be, but the draw is too much. And the denouement is always worth it because when she finally sees me I can see in her expression a conflict of choices. Should she run and leave the fish, should she run with the fish, or should she stay put? It is mostly the first, straight into the lake. It is never the second, but very occasionally, mostly when she is blinded by wind and rain, it is the third. Whether it is bravery or the simple fact that she does not see me, I cannot judge, but I'll do my bit, backing slowly away to leave her to it.

February did not disappoint. I always expect it to be the bleakest month of the year and this one proved to be equal to its fellows. There are no hints of spring, no harking back to the glories of autumn. Every day is the relentless winter. The landscape oozes cold and damp. But for all the deprivations, Kuschta was surviving well. As she passed her second birthday she was strong, fit and mistress of her domain. The valley was her place, providing all the shelter and food she needed as she ranged along its length. Naturally some days or weeks were better than others, but ultimately she only had herself to sustain in this solitary life. Strange though this might be, her whole life was predicated around the concept of being alone. She knew, of course, that she wasn't, because as food became scarcer the more other otters passed through the valley, but she'd do everything she could to avoid contact. She'd sniff at the spraints of uninvited visitors before trumping them with her mark. She'd constantly patrol what she regarded as the core of her territory to ward off incomers. She had no curiosity to meet others and they had no curiosity to meet her. It is a great system, barring one thing – procreation.

Early March was not appreciably different to the month it left behind but for one thing – a scent. Kuschta picked it out from the air; the particular odour of a very particular stranger. In itself that wasn't unusual, but this spoke to something deep inside her. Her first instinct was to double down on her territorial marking, but this time she paused, the particular dry muskiness percolating into a part of her brain that made her excited rather than fearful. Intrigued rather than dismissive. Curious, in fact. For a while she followed the trail, as it took her to higher ground, well away from anywhere she had been before. Eventually the unfamiliar landscape began to trouble her, so she turned back to home. As she slipped into the comforting embrace of her Badlands, bedding down in the dry shelter under one of the old brick culverts that harked back to the days when the water meadows operated, she was unsettled. For the first time since she had been abandoned by her mother, she felt the need to meet another otter. It was a need that was both visceral and strange, not least because it was that one particular otter. She wanted to ignore the urgings, but deep inside her something was changing that would never change back. She knew she must find this otter. But in truth, Mion was going to find her soon enough.

CHAPTER 4

ALONE BUT DETERMINED

Spring

Mion's job, if his life may be described in such terms, is that of a grand overseer, protector and inseminator. Whereas Kuschta could cover her territory in a day if pressed, Mion ranged far and wide, taking as much as a week to cover his. It might be tempting to describe the four or five females that lived under his purview as his harem, but that gives a lie to his active involvement, suggesting that Kuschta and her gender are supplicants. They aren't.

As the grand overseer, Mion had been aware of Kuschta from almost the very first day she arrived in the Wallop Valley, but chose to give her a wide berth, allowing her the freedom to mark out a home of her own. Two years her senior, he knew the valley better, confident of travelling further from the water than she was. With short cuts across fields at the junctions and bends of the river, he was able to dip in and out of her life unnoticed. Kuschta was too young for mating, so Mion was content to view her from afar until the time was right.

Like Kuschta, Mion tried to avoid conflict and interaction with other otters, but as the alpha male it was not always that simple. In truth, at somewhere between four and five years old, Mion was past his prime. Not in his dotage exactly, he had maybe one good year left, so, for now at least, young males mostly passed on and through the valley without challenging his authority. But it would not always be like that. There would come a time when Mion would have to fight for his dominance. For all their complex avoidance strategies, there is no way that otters are able to circumvent the Darwinian certainty of the survival of the fittest. There is some debate as to how much male otters fight. It is clear that they do fight, but nobody is exactly sure whether when two come head to head it comes down to a contest, or whether that is the choice of last resort. What is clear, however, is that it is simply not possible for two mature males to exist in harmony. One must give way to the other. The results will range from a mild confrontation to mortal injury.

Mion the protector comes in two guises. Firstly as pure and self-interested, keeping away competing males, and secondly, controlling the number of females so that the valley doesn't become overpopulated by them or their offspring. The females are generally no problem; the combination of Mion's presence, plus the territoriality of Kuschta, is enough to drive any newcomers away or at least keep them on the move. The males, on the other hand, are different. Of course the juvenile ones, recently cut adrift from the maternal holt, are easily cowed; they slink off at the slightest hint of his presence. But the older ones, now much the same size as Mion as they approach two years of age, have the same desires, and he stands in the way of those. So a fight must ensue. Be in no doubt that, while otter fights may be short, they are vicious – one in four will lead to a death.

The confrontation will happen fast; there are no pre-fight preenings or prancings. When Mion spots a rival that shows no sign of retreat the game is on. Fights usually seem to start out on land, and in an instant he runs headlong at the intruder, caterwauling in battle cry. His ears are pressed tight against his skull. The lips are curled back to expose the yellowed teeth, his face set in a crazed, almost rictus, grin. It is unlike anything you will see in an otter at any other time. The sharp incisors are ready to inflict the damage. Mion's tail, usually languid at rest, scythes from side to side, the rhythm increasing as the distance between the pair closes. Then suddenly, when face to face with their whiskers almost touching, the two pause. You think maybe that is the end, a sniff and then a dignified retreat by one or both. After all, everything you know tells you otters like to avoid confrontation. You assume the best.

And yes, they do sniff at each other's faces, but it is a courtesy, like boxers touching gloves in that moment before the fight begins. With high-pitched screams the two launch at each other. Teeth are the weapons of choice – the face, the feet and the vulnerable male organ area beneath the tail are the places to inflict wounds. It is a fight where one retreats and the other attacks. Fights don't last long. There is no soft, fatty tissue on an otter where wounds count for little. Driven home, those long backward-curved teeth, so good for subduing prey, go deep into the tissue. Legs, penises, cheeks, lips, scrotums, ears ... there is no single killer wound. Mion's aim is to inflict maximum fright and distress then retreat. He has no intention of shaking his opponent by its neck until he dies like a dog might kill a rabbit, or of tearing its throat out. All he wants to do is induce flight, and in a very short time it is over, the defeated fleeing to the river. Will the vanquished die? Well, the statistics tell us probably not, but for those that do it will probably be from festering injuries rather than outright damage.

Smell and sound are the senses by which otters order their lives; sight, poor as it is, comes a distant third. And so it was that the next time Kuschta knew Mion was close came with a call across The Badlands. To my ear otter calls are very distinct, unlike anything else that cuts through the stillness of an English night. They have a clarity and purpose to them – from an abandoned pup wailing for its mother to the happy chatter of a family on the move. Bring together two or more otters that want be together and the result is happy sounds; chittering, whickering, whistling, squeaking – they are all words that are commonly used to describe the chatter of otters. I don't know which word is best; for me 'eeking' is the word (not a proper one, I know) that describes the sound, both plaintive and informative at the same time. It's a bit like one otter saying to another, 'I'm over here but don't go far without me'.

It is generally believed that the European otter has six calls. They are not melodic, ranging through the scales like that of birdsong, but they are distinct and recognisable. I can't pretend to know them all, maybe two or three at most, but others can. Otter hunters of old spoke of 'love whistling', and it was with that call that Mion sought out Kuschta a few days later, in that way that otters do, calling from the distance, preparing her for his arrival, moving ever closer until he bounded into her life.

By rights Kuschta should have been scared of Mion; never in her life had she been this close to an otter so much larger than herself. But the calling and the scenting had prepared her for such a moment, so when he broke through the reeds into her clearing beside the brook she stood her ground as they came face to face. As with the fight, their heads meet, whiskers touching, but this time instead of a blood-curdling battle cry they exchange brief squeaks of pleasurable intro-duction before Kuschta bats Mion away with her paw, striking him on the head. The blow is not hard enough to cause any damage, her claws drawn in, but the message is clear: stand back, presume nothing. Mion takes his cue, retreating a few yards before quickly turning to face Kuschta again, pressing his body flat to the grass in an effort to appear unthreatening, his paddle swishing slowly from side to side. The tail is the messenger: all is calm, come to me when you are ready.

Crouching, as if to mimic his pose, Kuschta half slides, half crawls towards Mion, her belly rubbing the ground until the moment she gets within range when, with open jaws, she launches at his head. He's ready for the move, protecting his face with his paws whilst at the same time rolling over so Kuschta's head buries itself somewhere beneath his throat, allowing him to grasp her to his chest. Enraged at being pinned to him, she struggles to get free, the pair rolling over and over again, pawing and biting at every point of their bodies. The blur of fur is truly reminiscent of a pair of writhing snakes, but it has that playfulness of a kitten with a ball of wool. It looks bad, but you know it isn't.

Suddenly the two separate, the wrestle over for a moment. They take stock of each other, sniffing with weaving heads and exchanging calls, all the while facing each other less than a yard apart. Sometimes one bounds forward and the other retreats, maintaining the stand-off between them; this repeated dozens of times. Other times one is the supplicant, absorbing the attack by rolling onto the back, legs open and belly vulnerable, happy for the other to do the worst for a while before eventually calling time with a meaningful nip. Chase. Retreat. Engage. Chase. Retreat. Engage. That is the otter way of courtship. But it is an energy-sapping interaction and otters are not big on wasting energy. Soon Kuschta had had enough of Mion; not in the sense that she was displeased but rather he'd proved himself to be a worthy mate. After all, Kuschta will have only one, two or maybe three litters during her life at the very most. Why waste it on an inferior choice? Mion, on the other hand, is truly polygynous, ready to mate with more than one female, and as his territory covers those of many Kuschtas, he will do just that. If you've got a good gene pool spread it about, is the male otter mantra.

So do the newly bonded happy couple head off to the river side by side, frolicking in the water whilst they hunt for food together, companionably chattering on the bank to share a newly caught eel or fish? Not a bit of it – hunting remains a solitary activity and the two go their separate ways for the remainder of the night, Kuschta bedding down at her holt and Mion on a nearby couch, ready to do battle with any male that might pass this way. After all, a female on heat may attract many suitors, but the next nightfall sees the two alone still.

They come together, touching noses Inuit-style in a rare moment of unhurried intimacy that seals the relationship before they resume the wrestling and rolling, with bouts of grooming in between. They don't confine themselves to land, sometimes chasing each other into the river where they grip together, roiling up the water on the surface or disappearing out of sight, completely submerged for 15 or 20 seconds until two heads pop up for air. Occasionally the fights on land become so intense that the pair roll off the bank, tumbling into the river. They don't seem to notice the accidental transition from land to water, the pawing and clawing carrying on unabated.

Whether in the water or out, there is a definite change in Kuschta's demeanour. It seems that, having made her choice, she doesn't have the need to go head to head with Mion every time; sometimes she'll turn her back on him, allowing him to briefly mount her from behind. Other times she will let him roll her around. Gradually the play fighting gets less playful, Mion more dominant, the mock mounting more serious, more prolonged, with him biting into the scruff of her neck until she yelps with pain, breaking away.

The pairing up, if we can really call it that, is short – maybe four or five days at most. Actually, if you add up the total number of hours they spend together it won't be much more than a dozen, although they will remain in close proximity over that period. No doubt that famous otter preference for being alone has something to do with it, but really beyond the act of procreation neither needs the other very much and Kuschta will soon prove this in no uncertain terms. For Mion has to stay, but he also has to go. All the time Kuschta is ready to mate he has to keep near to ward off suitors, but in staying close the remainder of his territory is left unpatrolled, his other females unprotected. So, with the clock counting down, they were both more than ready by the third night.

That night was cold, the bright moonlight illuminating the frosted ground silver grey. Somewhere up the mist-wraithed river Mion was making his way back to Kuschta, who waits patiently beside her holt. As he comes closer she hears his regular calls marking down his progress, and she responds in kind. Coming into sight, his head is just a dark blob amidst the effervescence of his wake, but she knows it is him. His voice is enough and she whickers in welcome. Hauling himself onto the bank, Mion bobs and weaves towards her, proud to be there. The two touch noses in welcome, grab at each other with legs and paws, take a few tumbling rolls across the frozen grass, pop to their feet and side by side gambol back to the river. That was about as loving as it ever was – otter mating is fierce, frenetic and long. As the two hit the water they grab at each other, lying side by side, two-thirds of their bodies submerged as they nip and scrabble at each other. As the wetness turns their fur oil-slick black, the moonlight picks out the white-pink open wounds on their flanks left from the two days of 'play' fighting.

It seems that once the ritual has started they hardly dare let go of each other; they spin round together like a rolling log. They all but disappear under water, just a leg or a tail showing above the surface. Mouths open to gasp in breath, yellow teeth catching the moonlight. They look for all the world like a pair of writhing Olympic wrestlers trying to achieve two falls and a submission whilst all the while churning up the river, surrounding themselves in white foam. Then it becomes clear what is happening; Mion is trying to place himself behind Kuschta, but for the moment, by gripping him front-on, she is preventing this. Whether this is some form of elaborate foreplay or a genuine fear of what is to come, who knows, but her evasions cannot last forever. He is bigger, stronger and determined. Roll by roll, wrestle by wrestle, Mion gradually gets the upper hand, twisting himself over Kuschta's back until he finally grasps in his mouth a thick wad of her fur and flesh on the back of her neck – for the next twenty to thirty minutes he will not let go.

Kuschta might be captured but she is not yet subdued, struggling to free herself from Mion's grip, all the while holding her tail tight to her body to prevent him entering her. Up, down and across the river they go, their writhing bodies entwined like a two-headed sea serpent, Mion's head appearing first, shortly followed by Kustcha's when they come up for air. The noise is mostly from the thrashing of the water. Mion is mute, his mouth jammed tight into her flesh. Occasionally Kuschta screams in pain, but mostly they cough and splutter, expelling water and inhaling air like gasping asthma victims.

Eventually, whether by accident or design, they ground themselves on a gravel spit. Out of the water Mion's size and strength breaks Kuschta's resistance. She screams with a sustained, pained yikkering as he forces her to the ground, where they lie side by side. His front paws jam into the clefts of her thighs which gives him the purchase to come up and under her tail to enter her. Once he is in, they lie motionless for a few seconds, their tails stretched out behind them, as they and the tails spoon each other. But the respite is momentary. In rapid succession Mion's tail goes rigid, he squeezes Kuschta tight with his front paws, thrusting in further, scrabbling his rear paws at her behind as his whole body is overtaken by rapid spasms. He convulses as if fed by a pulsing electric current whilst Kuschta arches her head back, eyes closed, mouth wide open, teeth bared, continuing to emit that pained yikkering, but as more of a sob than a cry. Then all of a sudden Mion will go still, the pair motionless for a few seconds until the whole cycle repeats itself again. And again. And again. Just when you think it might be over, it isn't, the whole tableau repeating itself dozens of times.

Neither seems to take any pleasure from the mating; Kuschta is ever trying to escape, writhing on the gravel, half in, half out of the water. Mion hangs on to sustain his penetration, biting down hard on her neck for control when she makes his task too hard; Kuschta reacts with intense screams. Soon they are rolling over and over again. Blood patches appear on her sides where Mion's sharp claws have cut through the fur to her skin. At one point they come to rest against a log, Mion on his back with Kuschta lying with her back on his belly; she looks pitiful, gripping her little front paws together, as if in prayer, as she sustains another bout of his spasms. Then for no apparent reason, he flips her over so she is standing on all fours, him still mounted from behind. Kuschta seizes the opportunity, arching her back so Mion is lifted clear of the ground, carrying him on her back as she runs them both into the river.

In the water the coupling continues but you sense that the end is near. Still Mion hangs on, but the spasms are fewer, and when they come, they are just momentary. As before, they roll together, dipping beneath the surface, the two heads appearing in tandem for breath. But finally, after half an hour, the urgency has gone. Exhausted, Kuschta drags their entwined bodies into the shallows, where she violently shakes her entire frame, the two slipping apart as Mion finally releases his bite-grip from her neck, her scruff fur a bloody mess. Without a backward glance she heads for her holt, Mion making no attempt to follow, slowly swimming back upstream whence he had come. They are both done for the night.

Mion did visit Kuschta twice more, on the two following nights. On the first they mated again but it was less intense, less prolonged – both knew that the moment had passed and everything that nature required had been done already. On the second night Kuschta spat and screamed at Mion like a feral cat before he ever got close. The message was clear and it suited him just fine. He had ground to cover, territory to protect, other seeds to sow. His involvement as a father had begun and ended in the waning crescent of a lunar cycle. In the months and years to come he would mostly be an unseen presence that passed close in the night. One day their paths would cross again, but for now he was that colonial overseer, protecting his subjects and his dominion from unwanted intrusion. It would be tempting to paint Mion as benign, but his intentions would not always be good. Kuschta had driven him away with good reason.

Kuschta didn't have long to prepare for motherhood, as gestation is somewhere in the region of eight to nine weeks. She had a birthing holt to find, needed to feed well to build up her strength and a territory to secure. Strangely, the time of year mattered not a jot. By happenchance she would give birth around the turn of April into May, but generally English otters are totally oblivious to the time of year, producing young equally through all twelve months. Otters that live out in the extremities – the north of Scotland or the lashed east coast along the North Sea – are more circumspect, confining the breeding season to spring and summer. Extraordinarily, some otter species have the remarkable ability to suspend pregnancy for up to ten months, holding the fertilised embryo in abeyance until the time is right to resume the gestation. It is a trick I suspect some humans might envy.

All that said, March is the bleakest month in the otter calendar. The bodily reserves, depleted through the winter months, are at a low ebb and just finding enough food to stay alive is a task in itself. The Wallop Brook is of little help, running fast and full. The fish stocks have been hit hard, caught in a pincer movement of predators – herons, egrets, otters, pike – who all take their share, and the natural mortality post-spawning takes another one in five. The good-sized fish are in survival mode, hard to find in so much water as they hunker down in the darkest holes, giving Kuschta little clue as to their whereabouts. Out in The Badlands the food chain is running on empty. The eels are either long gone to sea or deep, deep down into the soil in search of warmth and food, way beyond the accidental excavation by a deer leg. The damp seeps into every tiny body; the voles, the moles, the mice – all the little creatures are hanging on for a hint of green, a sliver of sun, a few extra degrees on the scale before they venture out to start their new year.

Only the waterfowl seem immune. The swans, lifetime partners, so much the polar opposite of otters, are nest building with a vengeance. Most nights when Kuschta swims past their ever-growing doughnut of dead grasses and reeds, the pen sits proud in the middle, constantly picking and rearranging at the circle around her. The cob never strays far, tearing reeds from the bank, offering them up so she may carefully interlace the new ones into the weft of the nest. Eggs cannot be far off. The mallards are the soundtrack of this bleak landscape. All night, all day they go, in seemingly perpetual dispute with each other, noisily pairing and re-pairing. Suddenly in a flurry, for who knows why, the covey will take flight, doing a few circles high over the valley before they return, crashing down into the river. All proud at this pointless intermission, they puff themselves up, vigorously wagging their tail feathers as they re-arrange themselves yet again along the river, the ducks voicing that characteristic series of quacks that start piercingly loud and get progressively softer until dying away to nothing. The drake, on the other hand, just sort of cackles away. They seem happy enough, so who can begrudge them their fun?

Kuschta has no time for frivolity. She needs to find a holt in which to give birth, a private place where she and the pups will be undisturbed. For otters it is very much about finding rather than digging a holt from scratch – an abandoned badger sett or a giant rabbit warren is the perfect starting point. In the usual run of things you will not find either of these very close to a river; there's too much risk of flooding, and if they are really near the water they'd be tunnelling into the water table. And so it was that Kuschta sought out higher ground, and on a dry slope, under the canopy of a cluster of trees, she found her spot in a beech wood that looks down over the Wallop Brook.

For the first time in her adult life Kuschta was away from the sound of running water, the brook a thread of silver away in the valley below, a good half a mile distant. The old badger sett was long disused, largely destroyed when the tree under which it had been dug was blown over, exposing most of the tunnels to daylight. Sniffing around, she explored the crater that the upended tree root had created, peering into various cavities, rejecting some as too small, some as too big, others too exposed, until she found the perfect one in the corner, at the base of the root. Ivy and grass tumbled over the entrance, shielding it from view. Pushing inside, it was smaller than you might think – no more than a good-sized dog bed – with just about enough height for Kuschta to stand up. But with some work (otters are happy to do some digging) it would be a snug place to give birth, and the pups would be safe for the first two months of their lives until they ventured outside.

You might wonder why she ventured so far from the river,

as it seems out of character. Well, it is a maternal thing, something that drives European otters to seek out a natal home well away from their usual territory. Otters that choose to live along the coast regularly travel a mile or more inland; in Scandinavia three miles is common, and Scottish otters sometimes seek out an isolated loch many miles from any river catchment. All this makes Kuschta's choice look positively conservative. As to the reason, well, in all likelihood it is about protecting the infant pups. Soon after they are born Kuschta will have to leave them for a few hours each night whilst she goes in search of food. Doing that with a holt by the river would leave them in peril from their own kind; Mion is not beyond eating his offspring. Half a mile there and half a mile back is a mere bagatelle, a small price to pay for safety.

Enlarging the maternal holt is no great effort for Kuschta; otter limbs are strong and the palmate paws, so well suited to swimming, adapt easily to digging. Her tough claws rake away at the dry, chalky soil with enough length for purchase to free all but the biggest flints or stones. Between each of her five digits Kuschta has leathery webbing that allows her to use the paws like shovels, redistributing the spoil around the cavern. There is a plan. She wants a dark, safe space in the deep recess of her excavation, with just enough room to curl into for giving birth. In the front she pads down a level area, just back from the entrance where some light and sun might spill in. This is where the pups will play and feed for the first two months of their lives before they ever truly venture outside.

After a week she had a functional holt dug, but it was far from finished – it needed comfort, it needed warmth – it needed a soft lining. Kuschta had chosen well, for the perfect ingredients lay all around her in the woods; moss, dry grass, twigs and the occasional tuft of sheep wool. But it was a slow process as she could only gather what she could carry in her mouth. It took many journeys to lay down her birthing bed; the twigs weaved into this natural mattress to allow air to circulate. And when it was done she went to do the same to her holt by the river, which would become the family home when the pups were old enough to be moved.

Holt-building was all very well, but what Kuschta craved as she moved well into the second trimester of her pregnancy was food, and plenty of it. With each passing day her world was shrinking, and the desire to patrol what she would perceive as her territory became less important. In fact, she became almost invisible to others, as the scent of pregnant and lactating otters diminishes to almost nothing. It would be back soon enough, but no doubt for now this is some sort of protection mechanism that allows them to come and go in secret when they are at their most vulnerable. Secret or not to otters, the one place Kuschta was leaving her mark was my trout lake. Every night, without fail, she left her calling card – footprints on frosty grass or a muddy patch on the bank. But I rarely saw or heard her. Maybe the occasional splash, but I got the sense it was all business – arrive, hunt, eat, leave. Time and effort were clearly at a premium. But she wasn't invisible to me. A spraint beside her favoured alder stump. A few pink trout eggs blinking up from the grass in the pale morning sun. A bit of a raggedy tail or some fish scales – clearly every fish was being devoured in full. Nothing was going to waste. I guess you couldn't blame her; she was eating for maybe three, four or even five. I'd find out in time.

As the calendar ticked down, she slept through the days and most of the nights in her Badlands holt; only a search for food warranted any effort. Eventually she knew her time was close: she made one last trip to the trout lake and then, at dawn, headed up her path to the beech wood. The valley below was turning green. The hedgerows were heavy with the rank scent of the flowering hawthorn bushes, the woods flecked with buds of bright green leaves. Everywhere nature was procreating. Around the holt baby rabbits scattered at the sight of her approach. Above, birds were nesting, the young crying out for food.

Kuschta knew her time was measured in hours rather than days. In one last effort before the sun rose high in the sky she piled extra moss bedding into the holt. Sliding inside, she twisted around two or three times on her mattress to create a snug hollow and, insulated from the world, settled down to wait.

CHAPTER 5

AND THEN THERE WERE FIVE

Early days

The birth came quickly; by the time the sun started to drive way the chill of the night Kuschta was curled around four tiny grey balls of soft fur, protecting them from the cold in the curve of her belly, her legs and tail squeezing the infants tight into her warmth. For a while the five slept, oblivious to the world outside the holt – Kuschta exhausted, the pups doing what seemed natural in the dry, curled in the warm embrace of their mother.

Newly born otters remind me a little bit of moles; cylindrical bodies covered with fur that has the texture of cotton wool and which absorbs your touch. Grey is really not a fair description of the colour; the thought of that bland tone does the beauty of it an injustice. Maybe blue, in the same sense that greyhounds or whippets are classified by breeders as 'blue', is better. Or again, in the palette of the mink fur trade where sapphire or silver blue is better still. Regardless of the descriptor, the four contrast vividly against the dark sleek brown of their mother as they snuggle in.

Though weighing not much more than a chicken egg at birth, they are, as you might expect for a creature that will grow to over four feet in length at adulthood, relatively long in the body from the outset. At five inches, including the tail, one would fill the cupped palm of your hand, with the tail maybe stretching to your middle finger knuckle. As it lay there, you'd notice the oversized paws, shaped not unlike little human hands but mostly covered in that silver-blue fur, with pink toes and bright white claws. The flat head is genetic, an obvious otter characteristic. The eyes are closed tight. The wet nose, jet black, contrasts against the fur, as do the fleshy pink ears. The fur around the mouth is light brown, and from it sprouts, even from birth, a veritable forest of whiskers. These will act as 'eyes' for the pups for the first few weeks of their lives until their own eyes open. In fact, for an animal that will grow up so super-tough, otter pups are remarkably feeble; they can't see, they can't walk, they certainly can't swim. They have so little bodily strength that they can't even lift their head for the first four or five weeks – movement around the holt is largely confined to some shuffling around the nest.

Kuschta's sleep is ended by the pups nudging and pawing at her belly, trying to search out her four teats amidst the thickness of her fur. From the outset the pups are noisy, chirping like little birds, until their toothless mouths lock onto the teats, silencing them for the next quarter of an hour.

Laid out in a row along Kuschta's belly like furry piglets, they push and knead with head and paws to encourage the flow of milk as they suck, their tails wagging in apparent delight. Someone once described this movement as 'giving the impression of motorised matchsticks'; it is apt and true. Eventually the tails will stop wagging and the pups, both sated and exhausted, will let the teats slip from their blood-red mouths as they fall asleep, lying contentedly on their backs, paws crossed across their bellies like Friar Tuck, their little bodies involuntarily twitching from time to time. But sleep is not a luxury that will be afforded to Kuschta for a little while yet.

Leaving the holt with the pups unprotected is always a risk; feeble and defenceless, they are easy pickings, but Kuschta has no choice. The beech wood is certainly remote enough from the river to generally avoid the two most likely killers – dog otters and mink – but a passing fox or stoat is a possible danger and she can only hope that her scentless condition, which should continue for a little while longer, will cover her tracks and keep the den secret. The card Kuschta really has up her sleeve is to change her habits; mothering by night, hunting by day. And so it was that for those first few months of motherhood I started to see her at the lake more and more.

We sometimes underestimate the ability of creatures to make wise choices that ensure their survival. Moles move to higher ground to avoid winter floods. Swallows flock ready for departure to northern Africa. Eels slip from river to English Channel to catch the Atlantic current so they may ultimately breed off the Florida coast. And in her choice of a birthing holt so close to the lake (it is less than a mile) Kuschta had ensured that she'd be away from the den for as little time as possible whilst having an easy food supply more or less on her doorstep. Clearly at some basic level she has enough innate intelligence to weigh up options and act accordingly, which was pretty bad luck for the trout in the lake.

I have to confess I never actually saw Kuschta hunting in broad daylight on the lake; I suspect the mill is too busy a place for a shy otter, though there are plenty of accounts of otters hunting by day in remote Scottish lochs at a similar time in their lives. But as May gave way to June, morning and evening were the times I'd see her most, close to dawn or dusk. She'd arrive from downstream; it is pretty well a straight shot up the brook once she has come down the hill from the beech wood to the water. In theory, Kuschta could have cut her journey by travelling the diagonal rather than the two sides of a triangle, but I never saw her do that. Maybe moving over long stretches of open field is too unfamiliar or risky. She'd use the outflow stream from the lake to the brook as her sneaky way in, but she was never quite sneaky enough to arrive unannounced. The final barrier is a set of boards that control the lake height. For whatever reason, she was incapable of negotiating these in silence, grunting and panting as she hauled herself over, splashing into the lake on the other side.

In the stillness of morning (on reflection, her visits were mostly at dawn) the sound always pricked up my ears and I'd see her flat head emerge from behind the screen of the willow trees as she swam into view. She never seemed in any hurry, describing a steady arc as she circumnavigated the lake in search of a fish. Her body is mostly submerged, her head held high as if sniffing the air, whilst those long whiskers, dripping with water and catching the first glint of the morning sun, probe the depths for vibration. It doesn't look a comfortable posture, but it must be effective as she suddenly stops, her arched back emerging above the surface, Loch Ness monster-like, before she plunges head first under water. Success or failure is almost instant. Success is, naturally enough, a two- or three-pound trout snared in her mouth. Failure is panting and coughing. As a mother she seems less inclined to chase her prey than in her juvenile past, so she simply continues with her steady arc to find another. There are plenty.

During the early life of the pups, in the weeks until they were weaned, Kuschta had to eat the fish rather than carry it home. If it was still half-light she'd take up her favourite spot on the alder tree roots, but once the sun was up she preferred to hide out under the footbridge, eating every single scrap of the fish. I can't say I ever saw her bring a carcass into the mill wheel room, but she definitely hid out there occasionally when some activity or other blocked her immediate path to home, then she'd bolt as soon as she could. I think she pretty well worked out that she could do the thirty yards in an underwater downstream dash from the wheel to the weir in a matter of a few seconds. Once over the boards, she could relax away from habitation, cruising down the brook whilst hugging the cover of the bankside fringe until she was swallowed up into the woods below.

For Kuschta those early weeks fell into a steady routine; sleeping, feeding, cleaning and hunting. The pups, two dogs and two bitches, were not particularly demanding beyond their basic needs. They suckled at her every three or four hours, she groomed them with her tongue and left them to sleep the remainder of the time. She, as is particular to otters, became something of a clean freak, nudging the pups around the nest so she could eat up all their soft faeces. At two weeks things got a little uncomfortable as the first teeth appeared, but as the pups grew stronger and the warm air of spring wafted into the holt Kuschta could get some respite, stretching out on the lip to catch some sun as the pups slept.

Hunting was always her greatest concern; leaving the pups, finding the food and getting back in rapid-quick time was an effort. Kuschta could easily go as long as eighteen hours without food, but during that time she would have to feed the pups six times, so she'd become gradually drained. She'd fret as the night drew on, gnawed with hunger and impatient for the first fingers of light to slide through the branches of the trees. She'd mark the moment to leave when that first light glinted off the brook below, knowing that at pre-dawn the night creatures she feared most as predators for the pups would have headed for bed, whilst it was still too early for the daytime creatures to venture out. She had an hour, ninety minutes at most, to get there and back. Her downhill route to the water was now well trodden; a narrow path of padded ground, no more than the width of your foot, that weaved between the trees, stumps and fallen logs. As the beech wood ended she had a short dash across the grass meadow, dodging between the cowpats of the dairy cattle, who, early risers themselves, would stare balefully at her, all the while chewing on the cud, apparently incurious at her rapid arrival and departure. The grass gave way to the copse that bordered the river, at which point Kuschta gratefully gave herself up to the short slide that whizzed her into the brook. She paddled upstream towards the lake with purpose, only pausing to spraint once she had put a good distance between her and the holt. In days past, before the litter, she'd have stopped to groom, consider her options and maybe explore the pools and undercuts for fish, but with dependants, the oncoming sunrise and the desire for food, she drove herself on. At the lake she consumed the single fish as fast as she could; any other time she might have hung around for a while for another, but today, as soon as the last bite was swallowed, the few scraps cleaned up, she headed back whence she came.

Otters are no great lovers of hills; their rolling gait is ill-equipped for slopes. Going downhill, their rump seems to want to fly past their front with every stride, so much so that you expect a nose plant and forward tumble at any moment. Going up just looks a plain struggle; legs, so well designed for water, don't have the same kinetic strength to effortlessly drive the body uphill like, say, those of a hare – the progress is almost ugly and painful to watch. The flat gradient of the river valleys has spoilt them, but for now Kuschta has to face the daily climb to the beech wood. In a few weeks' time it will be a thing of the past as she will be moving the family, one by one, to the river holt. In the meantime, their safety trumps her athletic shortcomings, so she makes the journey because she has to.

Approaching the holt, Kuschta has no way of knowing whether the pups are unharmed during her time away; no sound percolates out from inside, but even awake and ready for milk the four have enough instinctive self-preservation to keep silent. Standing on the rim of the root bowl, Kuschta sniffs the air and checks the area around the entrance. A fox would have to scrabble in, spraying dirt and stones to grab the pups. A stoat is more nimble, but killing is more impor-tant than eating, so the ground would be strewn with bloodied, grey corpses. But none of this has occurred, and the moment Kuschta pushes her head inside the grassy curtain the pups chirp in delighted unison at her arrival, demanding a new meal, scrabbling and scratching at her until all four are lined up at the milk bar.

As July arrived, so came a change in the pups. Under the bright green canopy of the beech leaves they emerged, blinking, into the sunlight. Walking is no natural thing to infant otters; their first steps are as tentative and precarious as that of a human baby. But walk they do, unaided and unassisted by Kuschta, who looks on, content to watch out for danger. All day the family plays outside, tumbling and entwining with each other, constantly tugging at Kuschta's whiskers as they use the rest of her body like a playground ride. Occasionally one of the pups will head off a few yards, burrowing beneath a pile of dry leaves, head popping out from the top as if surprised by the outcome. Others take it in turns to chew at tree roots that poke out of the soil and that are gradually stripped of bark as the pups test out new teeth. But for the most part they all stay together as if connected by short pieces of string, their unity being their strength.

All these demands on her attention Kuschta takes in good part as the pups pass their two-month birthday, gradually weaning them onto pieces of trout that she carries back from the lake. At first they took some persuasion, so Kuschta chewed up the flesh, but as their permanent teeth grew through they became more accustomed to solid food, the demand for her milk gradually falling away as the four put on a spurt of growth. But the easy times could not last. Things were getting crowded in the holt in the wood; no longer was there room for all five in the snug chamber at the back. Kuschta was fine, but the two boys, now noticeably bigger than their sisters, pushed these siblings to the margins at night. A move to the bigger holt by the river, where they could learn to swim and hunt, would give some respite, but ultimately, and in not so long, Kuschta sensed that the five had to become four.

The competition for food was becoming tiresome, Kuschta having to bat away the males to give the girls a fair chance, and in the final weeks before the move Kuschta had to visit the lake twice daily to feed her ever-demanding brood; on the first trip she'd catch two fish, eating one for herself, returning home with the other. On the second she caught a fish and left with it immediately. Even if I didn't see her or any tell-tales, the fish in the lake told me everything I needed to know. No longer did the fish greet me with leaps and frolics as I did my morning rounds; my shadow and footfalls spooked them witless, scattering them around the lake like so many speeding torpedoes. After a week, maybe a fifth of their number had been culled. They had to be sensing that, with anglers by day and otters at any other time, there was no way to ensure survival.

Around this period the only dead fish I ever found abandoned on the bank, sometimes partially eaten, sometimes hardly touched – there seemed no particular pattern to it – were the really big specimens that had lurked in the darker recesses for a year or more. My guess is that Kuschta must have left these behind as maybe too heavy to carry, or perhaps she didn't have the time or appetite to eat them. At first I used to pick up the corpses to put them on the compost heap, but I have to confess that one day, annoyed with Kuschta for slaughtering yet another good-sized fish, I booted it into the lake, where it sank, settling on the bottom belly up, the white stomach shining out from the depths like a beacon.

Now dead fish in your lake really don't look very good; they don't do any real harm but, well, it looks bad, so I resolved to get that particular one out. That meant donning waders, finding a long stick and attaching a wire noose to the end. For whatever reason, I cannot remember why, I never got around to the task that day, but the following morning I was togged up and ready to go but for the life of me I couldn't see the dead fish. There was no way it could have washed away; the current in the lake isn't that strong and in any case it would have become lodged in the boards and grille at the outflow. Circling the lake, craning my head this way and that to cut out the surface glare, I scoured the depths. Nothing. And then out of the corner of my eye I caught a flash of silver/white, and it wasn't in the lake but rather on the alder roots. There, at Kuschta's favourite feeding spot, was the much-eaten body of that very same fish – clearly it was still too big to eat entirely, even at a second sitting. I know I'd been told that otters eat carrion, but this was the first time I'd seen it with my own eyes. It makes a certain amount of sense, at least in the context of a lake fed from the water of a cool chalkstream. In 24 hours the flesh would have decayed very little, the smell would have guided her to it and, spared the effort of hunting, well, why not take advantage? From that day onwards I have felt a little bit more kindly towards Kuschta when it comes to those uneaten fish she leaves on the bank; at least we have found a mutual way of stopping them going to waste.

Away from the water, the beech wood had become a happy home for the otters, more private and secret with every passing day as the countryside took on the soft bloom of summer. Weighted down by a full crop of leaves, the beech branches now hung low, creating a mottled green parasol over the hollow, keeping it mostly in the shade except for the dusty fingers of sunlight that crept through the gaps in the branches. The casual passer-by could easily go past and see nothing of the otters. That said, the approach is uninvitingly steep, peppered with stools of coppiced hazel that grow up then fall out, forcing anyone of any height to move forward, crouching, pushing the stems this way and that to create a pathway. Even if you negotiate the slope and the hazel, the rim of the hollow will most likely halt your progress, greeted as you will be by a phalanx of nettles that provide a stinging, protective cordon. Look closely and you'll most likely see some faint pathways through the barrier, bare pounded earth at the base of the stems. A fine way in for anything or anyone under a foot in height, you'd conclude, before turning away.

The hollow itself is festooned with foxgloves; like nettles, *digitalis* were the first plants to colonise the broken ground created by the upended tree root. The stems are tall, growing as high as six feet in the perfect conditions, with each column of blooms producing a million seeds – once they put down roots they are hard to dislodge. For the pups the pink, bell-shaped flowers are a source of daily fascination, as they track the legions of bumble bees that come and go to collect pollen. Occasionally one of the pups will try to leap up to grab a bee in mid-air, but it always ends in a tableau of ignominy as the bee travels on unmolested whilst the pup tumbles to the ground. More effectively they sniff at the tiny little black insects that make a home in the foxglove stems before licking them up as they scurry from cranny to cranny.

In fact the damp hollow holds all sorts of experimental foods for the pups; earthworms are a favourite, which usually ends in a brief tug of war between two siblings as the worm pings in two, the spoils shared. Beetles, bugs, snails, cater-pillars – well, really anything that moves is at first mauled, then licked, then mouthed. Not everything passes muster – the latter, caterpillars, are soon spat out as those long fine body hairs irritate the soft inside of the mouth. Some things are just too confounding; they watch the rabbits that share the hollow, with no real idea of what they are or how they fit into their landscape. Field mice treat the pups with positive disdain; quicker, faster and infinitely more nimble, they pass almost under the pups' noses as they go about their daily business. Sometimes a mouse will make a bad tactical mistake by going in range of Kuschta, who, given the mood, will leap on it, crunching it head first then body second with that wicked array of teeth before swallowing – the sharp bones of this little mammal make it too dangerous to share with young pups.

Now stronger and more aware of their surroundings, the pups are less of a worry to Kuschta, the urgency of her fishing forays diminishing with each passing day, even allowing her to revert to her nocturnal hunting habits. The drawback always is carrying back the fish, but given the choice between ravenous pups eating fish or suckling her milk she'd take the fish option every time. But for all the security of the hollow in the wood and the gentle cadence of their lives – sleep, hunt, feed, play and then sleep again – time is up. The pups need to learn how to swim and hunt; creepy-crawlies are all very well, but their future relies on bigger and more elusive prey.

The choice of moving day, or rather moving night, was not a considered decision by Kuschta – it was just that moment in time when the pull of the river finally outweighed that of the wood. The early days were drawing to a close. She'd been back to her holt in The Badlands many times in the intervening months, leaving her marks and checking for unwanted guests. Sniffing at the tunnels and entrance told her that maybe the odd stoat or rabbit had shown a passing interest, but that was about it, so on a sultry night in June the pups finally came into contact with the element that was to define their life henceforth – water.

It is over a mile to The Badlands from the beech wood, far too far for three-month-old otter pups who have barely travelled more than a handful of yards from home, let alone miles, to walk. Each would have to be carried in turn, so as soon as dusk fell she picked up the first and the biggest, one of the two brothers, Lutran, by the scruff of the neck, in exactly the same way that a cat or dog carries their young, pushed out through the screen of nettles and headed downhill. At the edge of the wood she followed her well-worn path through the tall grasses of the meadow. No longer was a quick dash across the open field required. The cattle were gone for now, the pasture tall and willowy just a few days shy of hay-making, so her route was all but invisible. At the edge of the brook the pair took a rest. She gently lowered Lutran into some soft grass, tenderly nuzzling and licking the back of his neck to ease the numbness of being carried. He is not light, somewhere close to two pounds, so on a par to a typical fish she'd been bringing home, but he is buoyant, which would make the final leg of the trip a bit easier.

Picking Lutran up again by the scruff of his neck, Kuschta used the slide to slither themselves into the river, disappearing underwater momentarily, then popping to the surface, Kuschta's neck arched upwards to keep Lutran's head above water as she pushed on downstream. That said, she didn't worry about the occasional dunking; it's a remarkable thing but infant otters, even those much younger than Lutran, are able to survive underwater for some considerable time. I read an account of an otter hunt that disturbed a mother otter in her holt with month-old pups. With the hounds whipped away and at bay (hunts made a point of calling off the pursuit if the otter was lactating or with pups), the huntsman watched her bring each of the three out in turn, swimming underwater with the tiny pup in her mouth until they surfaced, unharmed, some considerable distance later.

But with no such dangers Kuschta paddled them at a steady pace as they headed past the big white house, took a slight detour around the waterfall, the relic of a long-abandoned water pumping station, and silently passed along the brook. If you were watching from a distance you'd see nothing more than a black T-shape travelling mid-stream, breaking the surface as it went, followed by a silver wake. Under the road bridge, a splashy tumble in and out of the hatch pool beside the farmyard and on through the water meadow they went. The stream, the gradient a fraction less, meanders for a few hundred yards for the last leg of the journey, lined with hawthorn trees along both banks that create a dark tunnel, hiding everything from view. Popping out of the gloom into relative brightness from the moon, Kuschta can see The Badlands ahead. She stops swimming momentarily to walk across the shallow gravel of the forded cattle crossing, before ducking under the barbed-wire fence that trails in the water. She has arrived at the place she really calls home; the place she had always intended to raise her family until the forces of nature will split them asunder.

CHAPTER 6

AND THEN THERE WERE FOUR ...

Abandonment

By dawn she had all four pups safely in The Badlands; the holt was a considerable improvement on the beech wood, with three roomy chambers and three different entrances. Crafted by badgers, rabbits and the flood force of one particularly rainy winter, it was more of a labyrinth than anything the pups had known before. They scurried from one earthy room to another, surprising each other by popping out into the daylight in unexpected places. For the moment, as they dashed and tumbled, they ignored the water, though they were completely surrounded by it, the holt being on what is effectively a small pear-shaped island, the last fading remnants of the water-meadow engineering. At the tip, the stalk end if you like, there is a pool, fed by the brook that pours through the narrow entrance between two brick hatch walls, the oak hatch control in the middle long rotten away. The flow then splits in two, curving around either side of the slightly mounded island, no doubt created thus from the centuries-old spoil from the man-dug channels. One channel is today the full-on river; fast and bright with gently sloping banks to gravel shallows and that pellucid chalkstream water in which grows the waving green *ranunculus* weed. The other is more brackish, clogged with fallen branches, flag irises that bloom yellow in the spring and overhung with brambles. But for all the obstructions the water still forces its way through, albeit as more of a trickle than a torrent, until it curves back around the base of the island to re-join the main river.

The island, maybe half the size of a rugby field at most, is densely overgrown with willow trees that defy the order of any normal woodland. It looks a mess – dead stumps with tall new willows growing up from the rotten core. Half-decayed branches in various stage of collapse relying on the support of healthy stems before they eventually fall to the ground. The bright bark of the living contrasting with the rough, moss-covered bark of the dying. As I say, it is a mess, but a mess with a purpose for that is the way crack willows, *Salix fragilis*, grow, the mature tree shedding its branches outwards, knocking down surrounding trees to create a clearing from which the new tree will sprout, growing in the root ball of its predecessor. They say the bats of that imperious cricketer W.G. Grace were shaped from willow grown in the Wallop Valley. I have no idea whether this is true, but I do know crack willows can be dangerous – there is another clue in the Latin name, *fragilis*, meaning fragile or brittle. The trees groan and wheeze in the slightest breeze, the branches suddenly exploding, splitting lengthways with an ear-splitting whiplash 'crack', before crashing to the ground. It is no coincidence that cattle find other trees under which to shelter in a storm.

It defies logic that an otter has to learn to swim; dogs, horses, deer, cats, rabbits and even moles are able to paddle themselves to safety without any teaching. But apparently not a three-month-old otter pup. Have they not read their zoological classification? What part of semi-aquatic do they not comprehend? Clearly it must be the aquatic part, for learning to swim takes time, trial and effort. There are all sorts of stories about how otters learn to swim; mothers throwing them into the water whilst they watch from the bank, or, more sympathetically, picking each in turn by the scruff of the neck for a first dip before being returned to land. The cutest tale has to be of the mother that had her two pups mount on her back. She then swam into the river, gently submerging to leave them bobbing on the surface, unsupported, but keeping close until they had enough, then partially submerging herself again so they could climb back aboard for the trip back home.

For Kuschta it was none of the above; the pups learnt by her example, following her into the shallows, the gently sloping gravel providing a rather perfect beach. At first they'd dash in and out, spinning around, revelling in the novelty of water, charting deeper and deeper water by degrees until quite suddenly they are afloat. They show little fear of this new experience; in fact, quite the reverse; they seem to revel in what must seem an almost weightless state, floating not in the water, but on it, much like a beach ball might, but sadly with just as little control. Prey to the current, they don't have the body weight to hold themselves upright, their little legs splaying out sideways, so they whirl and turn out of control until Kuschta noses them back to solid ground. Maybe this buoyancy is some sort of protection mechanism for those first trials with water, so whether it is a symptom of their age, or the contact with water that prompts it, around this time the pup fur starts to change. It takes on a darker hue, losing that puppy fluffiness and becoming the more sleek, waterproofed, dual-layered ermine that makes the otter so well adapted to life in the water.

The pups soon discover that water is more than just play; they turn over stones, hunting whatever they find beneath – wriggling nymphs, snails or baby crayfish. Sometimes you'll see them poised on the edge of the warm shallows, like a cat staking out a rabbit burrow, waiting for a shoal of tiny fish, mostly minnows or sticklebacks, to swim their way. You know, for all the patience and guile (they show a surprising amount for youngsters), it is a doomed strategy. As the pup rushes headlong at the fish, spraying water this way and that, the fish scatter in an instant, leaving a creature that seems bemused by the turn of events, circling the gradually calming waters in search of something that isn't there but, at least in whatever passes for logical thought in that little otter mind, should be. Eventually the realisation dawns that the fish are gone, so the defeated hunter returns for another vigil. That one will end in exactly the same way as the first. As will the next, and the next, and … well, it has to be said otters are very slow learners when it comes to hunting. Here are a few comparisons to prove the point.

You might expect sympatric species, that is to say animals that exist in the same geographic area, to become independent of their parents at much the same rate. Badgers and foxes are classic examples of this, sharing as they do the Wallop Valley with the otters, experiencing all the same highs and lows of the changing of the seasons, the weather, fluctuations in the availability of food and so on. To misquote slightly a political slogan, they are all in it together, but extend the truth behind the soundbite, and it is apparently not so. As a badger or a fox you will part company with your parents at four months, but for otters that moment comes after a full year. Quite why this is so, nobody really knows; the difference is not just a few weeks or months, but three times longer. It is all the more confounding when you consider that badgers and otters are of the same mustelid family. Put in statistical terms, two months after their fox and badger colleagues are living independently, an otter will only catch 10 per cent of its own food, relying on the mother for the other 90 per cent. Only at eight months will it be regularly hunting solo, and even when reaching its first birthday the young otter will continue to spend half its hunting time with the mother. As we will discover, the *lutra* apron strings are long.

Kuschta does her best to teach the pups to hunt, but really they learn from trial, error and imitation. The closest to formal schooling is the injured fish game; cruel for the fish and frustrating for the spectator and something only she really tries out in the first week of immersion. It's pretty simple: Kuschta catches a small fish, say a fingerling trout of three or four inches that are abundant in the summer, gathers the pups and releases it in a shallow pool where there is little chance of escape. You have to feel sorry for the little stripy parr, partially injured for sure and most likely traumatised, as each pup takes turns to catch it. Under normal circumstances, injured or not, my money would be on the trout every time, as it darts across the shallows, evading the claws and teeth of all comers. But, like a Roman amphitheatre with lions and gladiators, there is no escaping the inevitable. However annoyingly hopeless the pups are – and they are hopeless, with reactions still months away from being anywhere near honed – the little fish gradually gets worn down to exhaustion. Occasionally you'll cheer him on as events take a hopeful turn and he evades the pups, almost making it out of the pool and into deeper water, but Kuschta flicks him back into the fray; I'm sure a Roman crowd would have booed her at this point. But nature takes no heed of spectators so on it goes until, by attrition rather than skill, the battered creature is caught and eaten by one or other of the pups.

As the pups approach their four-month birthday they are really starting to look like quarter-sized adults. The light chestnut fur, the wide-set eyes, the flat head, black nose and pale ears are unmistakably 'otter', though the overall torso is still slightly chubby, almost akin to a water vole. That lithe, elongated frame will come with adulthood. The tail, that whirring matchstick, is becoming a very distinct appendage and seems to hold an endless fascination for the pups them-selves, almost as if they can't believe such a wondrous thing belongs to them. Like a dog in the park, they spin around and around, chasing their own tail, or when that fun pales they nip at the tails of their siblings, which probably explains why all the tails start thick and hairy at the body, tapering to a slightly reddened bare end. As this stage the pups are not the assiduous groomers they will become in adulthood; life is too much fun to stop for such tedious activity. But Kuschta encourages them to groom at dawn before sleeping for the day. At first the pups lick each other as much as them-selves and it becomes yet another game, but soon the habit takes hold as they work out where the effort is best expended.

There is no doubt that the move to Willow Island has made Kuschta's life a great deal easier, even if she still has to harvest just about every last iota of food for the hungry pups – they daily expend far more energy than they ever manage to replace with self-caught food. Like most creatures, food is what drives the daily choices that otters make. When fully grown, the male needs something close to three pounds of protein daily – that is a bit more than 10 per cent of his body weight. Females need about two-thirds of that figure. I say protein rather than fish because otters are anything but catholic in what they eat. That's a lot of food by the standards of most river systems, so in guarding their territory otters are guarding their food source. Again, like most creatures, they will eat anything if driven by extreme hunger, but there is definitely a menu of preference for a river otter. Top of the list are eels – in particular, the front half more than the back half. You will sometimes see otters repeatedly whacking a freshly caught eel against a tree or the ground; nobody is exactly sure why they do this; playfulness or sheer exuberance is sometimes cited but that's a lot of effort for no real reward. In all probability they are trying to get rid of the thick, slimy and not very tasty coating from the skin of the eel.

Eels are the most bizarre of creatures. They start their lives in the Sargasso Sea, a thousand miles off the east coast of Florida, and when just a few months old, no bigger than your small finger, hitch a ride to Europe on the Gulf Stream. Reaching our coastline, they seek out a river, swim upstream, haul themselves onto land, slivering across grassland until they find a pond or damp ditch that they will call home for the next 10–20 years. They are not the quickest of growers; an inch a year is the average, dining on a diet of earthworms, insects and bugs. Eventually nature will speak, the desire to reproduce the siren call. During the warm summer nights they will throw off the shackles of their earthy domains, using a sense of smell to guide them back to the river. On the cycle of the moon they will head downstream to the ocean, hitch the return ride on the southern current back to the Sargasso Sea, where, almost spent, they have just enough energy to mate, lay eggs, then die.

Of all the foods Kuschta likes, in fact that any otter likes, for that matter, eels are undoubtedly the favourite. Packed with protein and fat, *Anguilla anguilla,* the European eel, the very same one of jellied eel fame, is, pound for pound, the most nutritious meal she will consume all year. In the summer Kuschta will catch eels in the river, but in the depth of winter, she scurries along her paths to seek out the freshly splayed opening created by the deer. Naturally the damp ditches are the perfect home for the eels; they densely populate the abandoned water meadows. In summer you will sometimes see them squiggling close to the surface in the wet mud, but in the winter months they bury deep for warmth and food, beyond the reach of Kuschta's fast claws and curved teeth. However, if the sharp hooves and long legs of the deer puncture deep into the wet morass, suddenly they become vulnerable. In the dark wetness you or I would struggle to distinguish anything, let alone an eel that is much the same colour and texture as the world it inhabits. But for Kuschta, perching in wait, she doesn't need to see anything. The eel world is in chaos as they squirm this way and that to readjust. As the water starts to fill the holes she waits for the tell-tale vibrations before plunging head first into the mire. In open water the eel stands a pretty good chance of escape. In this earthy ravine, none. Using her back legs for purchase Kuschta drives herself downwards towards the prey. It is a blackout. Zero visibility. But any movement of the eel, however tiny, is enough to guide her until her lips and nose push against the wriggling fish. Whether the eel has any idea what is happening to it I have no idea; I doubt it. After years of quiet isolation in the damp earth the sudden disruption must be utterly confusing.

The final capture is easy enough for Kuschta. These eels are no thicker than the hose of a petrol pump, so she grasps a section of the body in her jaws, her teeth pricking through the smooth skin, gripping without biting it in half. The eel twists in a hopeless attempt to escape. Kuschta vigorously shakes her head to subdue it as she uses her front paws to push herself backwards out of the earth into the fresh air.

It is these eels, just starting their summer migration to the sea, that become the staple diet of the otter family as they emerge from The Badlands' ditches or travel down from further upstream on the cycle of the moon. The pups have no chance of catching them, lacking their mother's experience and sharp hunting wits, but for Kuschta they are easy prey. In the summer, there's no following in the footsteps of deer; she simply lies in wait on an open section of the bank until the tell-tale rustle of grass gives one away as it heads for the water. Virtually sightless and almost certainly unaware, for the eel there is no escape as she dashes forward and seizes it by the neck. If she hasn't severed the spinal cord with that first bite, a few wild flails will finish it off. You'd be tempted to think otters are triumphant in such success. After all, it is no mean feat to connect opportunity with guile to pluck a meal almost from thin air. So is there any basking in success? A moment of admiration before the feast? No, there is none of it. When she takes her first bite of the eel, severing the head from the body, Kuschta is still breathing hard from the effort of capture, blinking away the mud from her eyes. She eats as fast as she can. There are no prizes in the animal kingdom for table manners. She'll stop when she can eat no more or the eel is consumed, and only then settle down to groom herself clean. Today, though, there is no time to enjoy the spoils of her kill; instead ,she trots back to the holt where she carefully bites it into four pieces for the pups.

After eels it is fish. Otters are not overly fussy about what type – trout, salmon, perch, chub, bream, carp – well, just about any freshwater species you care to name. Pike might be the one exception, as this is probably the only fish with the ability to fight back, inflicting damage to an otter, plus it is notorious for having tiny bones that might even grate on the palate of an otter, which as you will see has a famous ability to eat the most extraordinarily indigestible items in rapid-quick time. Along the coast otters are a bit more discriminating; predator fish like sea bass and mackerel they ignore. It is romantic to think this might be one hunter showing deference for another, but the truth is more practical: these fish are fast and hard to catch. Otters go for the low-hanging fruit, so to speak, those that drift about in the slack water or become marooned in sea pools – rocklings, dogfish, sea scorpions and the like. Small fish they will consume whole, but bigger fish they start eating from the head down, often leaving the body from the dorsal fin to the tail. There is only so much an otter can consume at one sitting, so it eats the most nutritious end (eels included) first – heart, lungs, liver and so on. Hence a really huge fish will be left almost intact except for a hole in the belly where the otter has buried inside for the vital organs.

There was a time when it was thought that otters didn't eat carrion (including dead fish), but, as I witnessed myself, they do. Of course that begs the question as to how an otter finds the half-eaten fish without vibration or movement. Otters actually have better eyesight underwater than in air; the pressure alters the dynamic of how the eye operates. Those long whiskers are also critical, picking up the vibrations of prey. But since Kuschta is searching out inert food in darkness there has to be something else happening. For years it was assumed that otters, along with most mammals, had no underwater sense of smell, but an inquisitive wildlife cameraman begged to differ. He placed a dead trout on the bed of a murky stream, set up night-vision cameras and sat back. Sure enough, within a few hours an otter appeared, making a beeline for the fish, which it promptly made off with. That was something of a eureka moment; the otter had found the fish effectively blindfolded by the dark and murk. But how? Watching back the footage in super-slow motion, you can see the otter expel a small air bubble from its nose which it promptly breathes back in. This is how otters smell underwater. It is still work in progress on the exact science, but this real-life experiment in south-west England dovetails with a laboratory trial in Nashville, USA, where water shrews sought out food in exactly the same manner.

The list of things otters will eat is as endless as there are things to eat: eggs, moles (they can swim), water voles, earthworms, rabbits, ducklings, frogs ... I could go on but you get the idea. Like any wild animal, for the otter the choice of food is often driven by the coming together of hunger and opportunity. Despite what you might think, the depths of winter, the months immediately either side of Christmas, are not the worst. In the river the trout and salmon are still in fine condition for spawning, with the added bonus of the occasional fecund female fish loaded with eggs. Up and down the banks the privations of winter are still yet to hit the otters' lower food chain: water voles, field mice, frogs, newts and even some late fledgling waterfowl are doing fine for a while. In her daily patrols of The Badlands Kuschta has the pick of them all, though her choices are determined largely by what is at hand and what is plentiful. In times of plenty she will specialise; eels are a prime example. In times of need, February and March, she will eat whatever is at hand. These are truly the danger months for otters, when nature's larder is most bare. The spent, spawned trout, gaunt from their efforts, are almost dormant and tough to find. The tiny fish, minnows, sticklebacks and bull heads, the breeding season still a few months away, are depleted daily. On the ground the cold damp does for the water voles, and the other tiny mammals are picked off by owls and raptors as keen for fresh food as any otter. It is a nasty cycle of descent and one that Kuschta does well to avoid. And avoid it she does, by doing what otters have done best for millennia − adapting to the gifts they are given, because of all the carnivores they really are one of the most highly specialised mammals.

There is a scientific way of expressing this: Kendall's coefficient of concordance. It sounds complicated, and if you see the mathematical formula in print it might well freak you out. The Kendall scale, which runs from zero to one, is used in all sorts of analysis, human and animal. Put at its simplest, ask a group of people an opinion on a particular subject. If they all had the same opinion the Kendall score would be one; if they all had different opinions then it would be zero. In the carnivore world otters score a 0.82 in their choice of food; that is to say, they will pick particular things to eat above all others, passing up 'ordinary' items of food in the expectation that they will be able to source their favoured food. This is unusual; most eat as they find; foxes at 0.44 on the scale are a good example, which makes them scavengers rather than selectors. Otters clearly have some higher ability to make choices based on future expectations.

Behind eels and fish, the number three on the Nether Wallop otter smörgåsbord of life is crayfish. They eat them with a passion. Oftentimes I will crunch my way over the bridges, so littered are they with crushed shells and claws. It always surprises me that otters show such a preference. I can only assume that crayfish are truly tasty and nutritious, as crunching your way through the hard, sharp casings must be a trial.

Crayfish really do look and act like small lobsters. They are not easy to spot as they creep along the river bed, well camouflaged as they are, the colour of the brown stones under which they crawl back when not on the prowl. Their segmented bodies are about four inches long, then add on another three more for the vicious pincer claws that extend out front. As with their sea relations, they are the dustbin divers, scavenging along the floor for anything dead or decaying – flesh or fauna – that comes their way. Sometimes in the spring and winter they will strike lucky, positioning themselves behind spawning fish to guzzle away at the freshly fertilised ova that tumble into their path. They themselves are not so free with their egg laying. They mate in the autumn, the female producing 200–400 eggs, which, again like those of a lobster, are carried under the tail until the spring. When they hatch, the infant crayfish continue to live on the underside of the tail through three moults, emerging bigger on each occasion, until, as miniature replicas of the parent, they head off on their own.

Negotiating the path from juvenile crayfish to a two-year-old adult is not without its dangers; small fish are partial to young crayfish, as are ducks and herons, but plenty make it all the way, so much so that there has been something of an explosion in the numbers of crayfish, though sadly of the wrong kind as the imported signal crayfish has eradicated the native white-clawed crayfish. Great news for otters who like to feast on these fecund interlopers, but less so on the natives, and a cautionary tale of unintended consequences. It started in the 1970s when news of a Swedish fad for importing signal crayfish from North America reached the UK. The idea, enthusiastically supported by the government, was that the crayfish would thrive in Swedish lakes, creating a profitable industry with plenty of employment. After a couple of years, with the arrivals doing well in their new Scandinavian home, the first shipments arrived in Britain, introduced to lakes and rivers up and down the country. As with their Swedish counterparts they did equally well, so what could possibly go wrong? Well, what nobody knew at the time was that signal crayfish carried the crayfish plague, an infectious disease that, though harmless to the carrier (and anything that eats it), was fatal to the native white-claws. At the time nobody really noticed; river ecology was a marginal discipline in that era so the invaders carried on their merry way until it was too late. Today the white-claws are all but wiped out.

I'd like to tell you that, despite this sorry tale, there is today a booming industry in crayfish across northern Europe. A natural harvest of these tasty crustaceans that has bought wealth and employment to places where it would not have otherwise existed. But the clue is on the label of any tub of supermarket-bought crayfish tails: Product of Thailand. Sadly, the project failed, not because the crayfish didn't grow but because the processing was too expensive and forty years ago the rocket and crayfish sandwich was still yet to catch on. All that said, at a time when just about everything was mitigating against the revival of the otters, the upsurge in the crayfish population was a timely boost, as the newcomers are about ten times more populous than their predecessors ever were, and bigger too.

It is not just Kuschta and the pups that like them; for Irish otters, in places where crayfish are plentiful, they can account for half the daily diet. Otters that live along the coastline don't have the same predilection for crabs – they will eat them, but not a lot. That said, when times are hard crustaceans will always be a food option, with otters not beyond tearing into lobster pots and making off with the occupants.

For all the abundance of food at this time, a clock is ticking in Kuschta's head. Dividing food five ways in times of plenty is all very well, but she knows that the pickings of the previous winter were lean enough for her alone; add in two strapping young males, plus two females not far behind, and suddenly the equation looks unbalanced. The truth is that a litter of four makes for a big otter family; twos and threes are more common, and given that the Wallop Brook is a small stream with limited food, something has to give. And that give has to come in the abandonment of one of the pups to die, an act which must come very soon whilst the pups are still reliant on Kuschta for food.

The thing about living at Nether Wallop Mill is the constant noise of life by the water – it becomes the soundtrack of your existence, and, when you leave, the silence of other homes seems louder than no sound at all. After all these years I absorb the noises rather than hear them. The pounding rush of water through the millrace. The creaking turns of the water wheel, which has a habit of suddenly groaning as some obstruction halts its progress for a second, jumping forward as the force of the water moves it on again, the steel blades grinding against the wall as it momentarily gets out of kilter. The weirs tinkle away. The streams, all three of them, gurgle, cough and sigh. Above all this rises to the ear the evidence of the night-time crew. The hoots of the lonesome owl that cut through the dark. The piercing scream of a rabbit impaled by a stoat. The occasional splash of a trout that, for reasons I never understand, chooses to make a leap in the lake. All these are regular and normal, wafting through the open windows of a warm summer night.

After a while I even took to placing Kuschta's perambulations in the 'that's life by a river' column, no longer drawn from my bed to the balcony to watch her night-time progress. Lying there in the dark, I could trace her by sound alone: the arrival splash over the weir, the plunging pursuit of trout around the lake, the coughs and wheezes more audible with each failed dive. Then the silence which told me of success. I knew she'd be at the base of the alder tucking into her fish. A while later another small splash and she was done with eating, her fast strong swimming reverberating across the lake as she headed for home, and with one last flop over the weir she'd be gone. But when she brought the pups it was a very different story.

For all the things they did in their time at the mill my abiding memory of Kuschta's family is the noise; they arrived with noise, stayed with noise and left with noise. I wouldn't go so far as to say it was deafening, but at the height of the summer occupation I used to bury my head under the pillow by the third or fourth hour. It all soon fell into a regular pattern, with hardly a night going by that they didn't appear. Dusk seemed to be the trigger, sending them on the move from The Badlands, so that by the time they came close it was turning dark. That said, they were never the first arrivals; that honour fell to the bats who came out in force as twilight faded, the last reflection of the sun's rays disappearing from the sky. Swooping around the trees and circling above the lake, their task was more urgent. They needed to exploit the last heat from the day in the constant search for the thousands of insects that each must consume daily. Soon hunting for them will be over as the relative cool sends their prey to bed, so they take full advantage whilst they can.

Otters, of course, have no such worries – warm or cool, it matters not a bit to them. Darkness, or whatever approximates for it during the shortest nights of the year, is their true friend, with few other nocturnal creatures to challenge their turf. There are, of course, badgers, foxes, stoats and weasels, not to mention the occasional domestic cat, plying their own trade at the dead of night, but they know better than to mess with the otters who, for a few hours, make the valley their own. Discretion is nowhere in the lexicon of an otter family that is secure in its own territory; from a long, long way I can track their distant arrival by the eeking cries that travel across the meadows. This is not a cry of distress or panic, it is the companionable, delightful sound of the family chirruping to each other as they make their way upriver. The sound really does build, and build and build the closer they get. Eek … pause for three seconds … another responds … eek … another three seconds and another responds. So it goes on, like a moving echo chamber, until they arrive, tumbling over the weir, splashing into the lake.

During the first nights at the lake I never really saw much attempt at hunting from the pups, but they seemed to show no fear of the lake, by some margin the biggest body of water they would have ever encountered, as they followed in Kuschta's wake as she hunted. I never quite worked it out, whether the pups tired quickly or whether they were preventing her catching fish, but in those early visits she'd soon swing back, ushering them onto the island, giving them a helping nudge from water to dry ground where they would gather together to watch her in action. Sometimes I'd use a torch to pick them out, eight eyes in a line, shining back at me, white bright like the cats' eyes on a highway. Strangely, the beam never seemed to bother them or Kuschta; like her pups, she'd turn her head to follow the track of the light back in my direction before carrying on with the hunt.

Success, in the form of a fish, usually came quickly. At first Kuschta used to join the pups on the island for a family feast, but after a few days she headed for the bank by way of a change, carrying herself and the fish onto the grass where she'd place it at her feet, whickering in the direction of the pups. It took them a while to work this out; at first they thought it some sort of game, eeking back, reaching a crescendo of noise on the island. But soon Kuschta wearied; they didn't seem to be taking the hint so she simply settled down to eat. After a while the pups went quiet. You could see that somehow a message had got through. They perched on the edge, front paws gripping the turf, noses pointed downwards, almost touching the surface, their bodies flexing and twitching with indecision. All the quartet seemed to know that food lay just a dive and a swim away, but none of them had the gumption to take the plunge.

Eventually one of them, I couldn't tell which, seemed more

to overreach than actually dive, tumbling into the lake, already swimming for the far shore as it popped to the surface. Plop, plop, plop, the other three followed in rapid succession, all wriggling up onto the bank to share the booty with Kuschta. It is hard to judge these things, but I had a feeling that Kuschta was, for the first time, proud of her offspring. Some test, however small, had been passed, so she sat back grooming herself, surveying the scene whilst they made an unholy mess of the carcass. After a little while she slipped unnoticed into the lake to capture another fish, which they all eventually shared. It was getting to that point that they were definitely a two-fish family.

As I have said, I can't pretend to have the ear that recognises what are supposed to be the six inflections of the otter cry, but I know distress when I hear it. When it came, it came bad, cutting through the night, so very different to anything I had ever heard before. As ever, the family chatter echoed up the brook, that noisy precursor to their arrival. But this time it was different. Yes, there were the four voices of happy chatter, but there was a fifth trailing in the wake. Plaintive. Helpless. When they arrived it was clear there was some sort of schism. One of the pups had been ostracised, keeping a distance from the others, whilst all the time emitting that three-second pulse of wailing that sometimes pitched to a scream.

It took a while but it was soon apparent to me that it was one of the brothers; they were by now noticeably bigger than their sisters. At first I thought he was in pain, injured maybe, but he was able to run and swim without any apparent difficulty. His distress was entirely about his exclusion. The family had turned its collective back on him and almost the saddest thing was how total it was. It was as if he didn't exist. However wild or extreme his cries became, his siblings paid him absolutely no heed. He was just a shadow. At feeding time, as the four settled in, he went silent, slowly moving along the bank towards them, trying to blindside Kuschta. But blindside or no blindside, he really stood no chance, Kuschta turning at him as he came close, spitting and barking in the most venomous manner, sending him scurrying off. Later he tried again but met with worse as Kuschta rushed at him, lashing out with her front paws whilst the others never even looked up from eating. Cowering a little way off, he kept up the pained cries, but they were more occasional, as if he knew no one was listening.

He didn't move until the family were done, heading for home. As they left he made a beeline for the pounded grass, licking at the ground, trying to find some nutrition. I couldn't tell from a distance what pickings were left, but on past evidence I doubt very little. Certainly nothing to sustain him for any length of time. And then as the others disappeared into the distance, accompanied by that inevitable chorus, he stopped scavenging, sitting up to follow the sound. He paused, but not for very long. The pull of the family, however unwelcoming they might be, was more than that of the little food he could find so he ran to catch up, resuming that begging cry for recognition as he went.

However much the abandoned one sought reconciliation, it was clearly of no use, for over the next two nights his pariah status altered not at all. Why Kuschta had selected him from the four for a slow death by starvation and distress I have no idea. There was no reason to think he was the runt of the litter; he simply drew the short straw. Postmortems on the corpses of the abandoned pups show them, aside from the immediate cause of death, to be in good health, with no congenital abnormalities. I'd like to report that what I saw was an isolated instance, but it isn't; all the literature and research seems to suggest that this is really a very common occurrence. Otters along the Scottish coastline have even been seen to carry very young pups to rocks hundreds of yards out to sea where they are left to die. Why they do this is largely conjecture, but it is more than just a heightened survival of the fittest process. It is too deliberate and purposeful for that. Ultimately it has to be a brutal form of population control.

By the fourth night I had had enough of listening and observing from a distance, so on one of those long, warm summer nights when it never really gets dark I headed outside. You must remember that at this point I knew nothing about the concept of abandonment; the whole scenario just seemed inexplicable to me. What I hoped to achieve I have absolutely no idea, but as I walked along the edge of the lake I could see Kuschta and the brood surrounding a dead fish, with the waif a way off, as had become the norm. Greed often trumps fear in wild animals. It never ceases to amaze me how close I could get to the family when they were in full devour-ment, but eventually suspicion of me prevailed over hunger as they slid from grass to water, cruising just out of reach, watching my every move. I guess they knew by now they could always swim faster and better than me.

The four at bay, I confidently expected the abandoned one to fall on the food. Maybe that was my plan. But he didn't. He ran up to me, stopping at my feet, rose up on his hind legs, opened his mouth, drew back his lips over his yellow teeth and, whilst clawing the air with his front paws, screamed at me without pause. It was horrible. It was rabid. It simply didn't stop whilst I stood there. Had he been any size, I would have been truly terrified; I almost thought he was going to attack me anyway. I had absolutely no idea what to do. Pick him up? He seemed beyond that point. In the end I left him, turning my back on him, and headed indoors with his continued screeching following my every step until I closed the door to the house.

The following morning I went about my rounds, circling the lake and walking down the brook. Halfway to The Badlands, lying on the edge of the grass path that we mow along the bank, was a stretched-out body. Our little guy was dead. The emaciated frame was stiff with rigor mortis, the pale fur of the underbelly catching the sun. The birds had already pecked out the eyes, but other than that there seemed to be no injury; he had simply laid down and died. If I'd have known then what I know now, I sometimes wonder if on that last night I should have scooped him up, fostering an orphan. It would have been the humane thing to do. But otters are not human and their choices are best respected.

CHAPTER 7

A COUNTRY PLAYGROUND

The family summer

If the transition from five to four caused the otters any trauma it didn't show; life went on as before with, it seemed, only me wondering about the how and the why. That is the thing about wild animals; they don't register a loss like we do. In fact, almost the reverse. For the sisters it was one fewer dominant brother. For the surviving brother, he moved a notch up the pecking order. For Kuschta it made life that bit easier. For them all it meant more food and less competition in the gathering of it. Otters are an unsentimental lot and, as a grumpy teenager might intone, act as if 'what's not to like?' when the litter was reduced by one.

The daily, or should I really say nightly, routine changed very little in the early weeks after the abandonment, but as the pups grew it became easier to tell them apart, individual differences gradually emerging. Lutran, the sole surviving brother, was always easy to pick out. In terms of body length he wasn't that different to his sisters, but he had a greater bulk about him which he carried with less grace than the other three. We like to call otters lithe, it sums them up well, but he was a long way from that, still awkward and clumsy, and he would remain that way for a while to come. Where the others slipped into the water, he more lumped in – I didn't even have to look to know it was him, such was the splash and subsequent wave.

The sisters, Willow and Wisp, were harder to distinguish between, but they were both busier than Lutran – more inquisitive but less prone to sudden actions. Where he bolted his food, they took their time. When hunting, he'd launch himself at the prey with no forethought, whilst they'd steady themselves before making the attack. In time I'd get to know which of the sisters was which with ease, but for now it was always something of a guess. Inevitably I rarely saw them all in full daylight or completely dry, so the true colour of their pelage was harder to know – they mostly looked slick black, fresh from the water or scooting around in the dark. So one morning when I caught them napping (for some reason they had chosen to hang around until well after sunrise) it was something of a shock to see that the grey fur had been supplanted with a coat more light chestnut, the soft baby pelt gone, their body covering more resembling Kuschta's with those long, thick outer hairs over the waterproof coat beneath.

As travelling companions I suspect they must have tried Kuschta's patience to the limit. For the three of them, keeping up with her was a relentless competition for they vied to be the lead pup as they trotted, ran or swam in a V formation behind her. It clearly never occurs to young otters that energy is best conserved. No sooner has one gained the front spot than the one behind gradually speeds up to go past to take the lead, soon to be replaced by the third, who in turn will be replaced by the one that was originally the frontrunner. It is truly an ever-revolving trio and actually quite tiring to watch, so relentless is the competition between them for the lead position. Occasionally one tries a really bold move, coming up alongside mother with thoughts of passing, but a quick sideways glance from Kuschta will put paid to that, the sudden slowing causing a minor traffic pile-up behind.

Though it matters not the season in which a litter is born, by my reckoning Willow, Lutran and Wisp had been dealt a pretty kind hand. If I ever have to come back in another life, being a young otter amongst the luxuriant summer beauty of an English river valley would come pretty high on the list. For every yard of the way, from the tiny beginnings where the brook emerges from the ground to its confluence with the River Test seven miles later, the Wallop Brook is a truly magical place. Every twist and turn is cloaked in an abundance of meadow plants – hemp agrimony, meadowsweet, fleabane, comfrey, marsh marigold – the names just trip off the tongue, evoking images of bucolic beauty. The truth is they look every bit as wonderful as they sound.

If you ever have the chance, do take a moment to pause to look along a river bank to see what Mother Nature does quite perfectly. She achieves with little apparent effort what a professional gardener might spend a lifetime trying to achieve in a herbaceous border. Those drifts of colour with variations in height and texture that catch the eye, where both the individual plants and the overall design draw you in, where there is something to catch the eye from the lowest to the highest points, with plenty in between. The water margin is pecked with the vivid yellow of the marsh marigold, the flowers like oversized buttercups with leaves that so much resemble watercress. The pendulous white and purple comfrey flowers droop out over the water, surrounded in the warmth of the day by the regular drone of bumblebees that fly with unerring accuracy into the heart of the tubular blooms. For a few seconds their busy little red rears pulsate, poking out of the end of the flower as they hoover up a batch of pollen. Then they back out, falling for a moment through the air until their wings, a black blur to the eye, catch enough lift to send them on their way. But treat the comfrey plant with care; it has sharp, hairy stems that chafe the skin.

Even the gentlest of breezes moves the tall meadowsweet, the soft white flowers like so many lambs' tails waving along the bank. Its other name, the lady of the meadow, gives a clue to the delicate nature of the heads that contrast with the stiff purple and reddish stems. Of all the plants along the river, it is probably the only one that can claim a place in modern medical history; salicylic acid, the basis for aspirin, was first drawn from this plant. Equally tall, but really the one with the dominant hue, is the purple loosestrife. It flowers all summer with fifteen or twenty spikes of magenta blooms growing up from a single plant, which thrives with roots that push down deep into the damp river bank. If comfrey is the home for bees, then loosestrife is the home for all kinds of butterflies, but it is tortoiseshells, with those black-flecked orange wings with tapering blue edges, that dominate in this particular part of the world, gadding around from sun up to sun down, feasting on the nectar.

Of course, not everything that grows in the margins and on the bank is full of colour. The tall, spear-shaped reeds of the flag iris look very ordinary now, the yellow irises a distant memory from way back in spring. The hemlock is another plant on the wane at this time of year, the weight of the drying seed pods snapping the overgrown stems to droop down into the water, creating the perfect ladder for drinking insects to climb. Giant clumps of water dock are muscling in. The dullest plant with uniformly drab but huge, dark green leaves, it stakes out its territory, dominating where it takes hold. As a late developer it seemingly waits until the others have shot their bolt, growing on, up and above the rest, saving its spikes of tiny, pinky-green flowers until late summer.

All this makes for a very secret river. Walk along the ridgeway above the valley and you will see why – it is a great vantage point from which to survey the panorama of the countryside, a place to understand how a countryside evolves. Whenever I traverse the ridge I feel a sense of awe, a certainty that I'm travelling in footsteps that have trod this route for thousands of years, way back to the time before even Stonehenge, just twelve miles distant, was built. Today the track is a road, but a hundred years ago you'd have been on a rutted trail, a mixture of dirt, chalk and flints, pounded by the hooves of millions of sheep that would have been driven along this way in centuries past. It is truly an ancient route. This region is dotted with Iron Age forts on the surrounding hills; the structures are long gone but the earthworks are still visible. These elevated tracks were once the connecting highways of Stone Age man.

If you raise your eyes to the horizon you'd pretty well see exactly what our Neanderthal ancestors would have seen: the gentle downs stretching way into the distance, counterpointed by the occasional higher hills on which those forts were built. In the clear, sharp cold of a winter morning the view just goes on and on; in the summer it is shaded out by a slight haze. The downland that rolls away on the slope below you is certainly different now, but only recently so. Just three or four generations ago ancient man would have been looking at the same landscape as us, even though separated by millennia – close-cropped rough turf dotted with stunted hawthorn bushes with bright, white pockmarks in between. The latter would be intriguing. You might be pardoned for thinking that army shells from nearby Salisbury Plain had gone astray, the white holes and surrounding chalk debris resembling the landing places of so many exploding mortar bombs.

The truth is more mundane, but more appropriate. The holes and debris are the diggings of rabbits who would have proliferated in their millions. It was they, not the sheep, who kept the grass so close-cropped, and they would eat anything else that had the temerity to grow, hence the lack of woodland. For rabbits the chalk downs were heaven sent, perfect for dry warrens, and the open space made it hard for predators to approach. That is, of course, with the exception of man, who exploited the relentless breeding capacity of rabbits to the degree that their meat provided a staple part of the local diet. But the downlands, and the rabbits to boot, largely met their match with the advent of the Second World War. In the drive for agricultural production, this virgin land, never before touched by man, was turned by the plough. Grass was replaced by wheat, the rabbits driven to the margins along the headlands of the newly created fields and amongst the hedges that divided them.

Whether this change is good or bad is hard to say. I've seen both and I have to confess I like the regimented beauty of crops. In the winter the bare earth is a swirl of patterns where the harrows and seed drills have done their preparations; the look is of an earthy Bridget Riley painting writ large, making an optical illusion of the ordinary. In the summer the swaying corn paints its own pictures as gusts of wind push the fields this way and that, creating the effect of a gently pulsating golden ocean. Where the crops end, the verdant ribbon in which the river hides begins; in the height of summer you'd barely know it was there, down in the valley. Back in the winter, with the trees naked and the ground bare, you can trace the bright progress of the brook for a good mile or more in each direction from the ridgeway, but now you just catch the occasional glimpse where the water meadows briefly open up. Somewhere down there three young otters, coming up to their five-month birthday, are trying to make sense of a complicated world with the benign help of their mother.

I am not always sure that benign is the best word to describe how Kuschta's relationship with her brood was evolving – warm-hearted, tender, compassionate – these suggest something different to the truth. The pups were growing fast, Lutran in particular. They could swim. They could run. They had a basic understanding of what was good and bad in the otter world. They were way past depending on Kuschta for milk or warmth. But, and this is a big but, they were completely unable to hunt down enough food to survive alone. If Kuschta was killed, maybe run over on the road (incidentally, this is by far the commonest way for otters to die aside from natural causes), the litter would be dead from starvation within a week. This is common enough. In maybe a couple of months Lutran might have a small chance of survival, but he'd do it alone. The mother is the glue that keeps the siblings together. Without her they would split, to survive or die alone. So Kuschta, knowing this, was trying to become more of the overseer, guiding the pups to the places they needed to be whilst going about her daily life with the troupe in tow, letting her actions be her lessons. She was to be the grand conductor directing her fumbling acolytes to achieve success through trial, error and constant practice so she could finally vacate the rostrum unnoticed.

In the sanctuary of The Badlands, hidden from the world by the profusion of growth, Kuschta felt safe to let the pups roam a short way from the holt during the warm summer days. Letting them explore, she took to sunning herself from a look-out on the top of a tall stack of willow timber left, for whatever reason, by the woodsmen who pollard the trees every few years. Crack willows are all very well in theory, but in practice they need to be managed, hence the thick gnarled trunks, some hundreds of years old, with fresh growth shooting from the crown. My theory is that Kuschta liked the pile as much for the peace it gave her as for the warmth and vantage point; it would be a while before the pups were nimble enough to make it to the top.

Below her the pups are a blur of action; when awake, they really don't seem to have an off switch. It would be tempting to call it play, but surely there has to be some higher purpose? They writhe, tumble and splash with each other in the shallows, then an instant later flip themselves onto the bank, rushing at each other to meet head on, absorbing the impact with paws outstretched before clasping together to roll over and over, entwined. Then they separate, bounce up and do the same thing again. And again. And again, rushing hither and thither. It is actually quite exhausting to watch. As to that higher purpose, one can only assume that the agility and dexterity required to catch elusive prey are learnt this way.

Of course, as we know, fish and eels are the supper of choice, but for Lutran, Willow and Wisp these are a distant dream unless provided by Kuschta. For now they are, quite literally, hunting at the bottom of the food chain. Some things are easily found; the pups soon become adept at turning over stones to reveal a whole host of what might be termed creepy-crawlies but which are far more important than that loose term might imply. Freshwater shrimps, *Gammarus pulex*, which look very much like their sea cousins, but writ small – you could probably get half a dozen to fit on your thumbnail with ease – are to be found everywhere. If you gathered the entire population from a stream even as tiny as the Wallop Brook it would run to millions rather than thousands. The diminutive size of these crustaceans belies their importance; they are one of the building blocks of the food chain, eaten by any and every one, including the pups, who suck and lick them up in the early weeks until they discover better nutrition that is more easily found.

There are tougher nuts to crack as they scrabble in the river bed; freshwater snails don't quite come in that category, easily caught and crushed, even by juvenile teeth. Mussels, dislodged from the mud, are more of a challenge and are chewed on like a dog might a tiny bone; no doubt a good strengthening process for the jaw and mouth and clearly a certain amount of fun as they paw them around, flicking them up then knocking them back into the river in what seems to be a rather one-sided game of hide and seek. The biggest prize of all fell to Wisp, who was the first to capture a crayfish, which always seems to me to be one of the most unlikely denizens of any river, let alone a small chalkstream like the Wallop Brook. But sure enough there is a thriving population (though not native crayfish) that provide a huge source of protein and on this occasion a certain amount of bafflement and pain.

If catching crayfish was sometimes hard, eating them was not without its problems. Make no mistake, crayfish are no pushover; they wave those front claws in the air with intent, opening and closing the pincers to ward off all comers. How Wisp had found her first crayfish I have no idea; I suspect she just saw it shuffling along the shallows and with some luck grabbed it by the tail, dragging it to the shore. The protesting crayfish reared itself up as they do, whilst the ever-so-innocent Wisp chose to take an exploratory sniff at her new friend, only to be promptly nipped on the face. Whickering with pain, she lashed out at her attacker, knocking it further up the bank with her front paw before leaping at it, this time pressing it to the ground with both paws.

Now otters actually have a tried-and-tested method of dealing with crayfish; after all, they need one. At certain times of the year they are a critical source of food, recorded as high as 80 per cent of the daily intake for Irish otters living in the west-coast rivers and lochs of the Republic. It took me a while to work out what the otters were doing, because they'd sit up on their haunches, almost meerkat-like, clutching the crayfish in their front paws. This seemed to me at best ill-advised. Surely they, like Wisp, would be nipped at close quarters? Furthermore, there seemed to be no rationale as to which way up they held the crayfish – sometimes the squirming crayfish was head up, other times tail up. The only certainty was that it was never sideways. Eventually I worked it out; the otter manipulates the crayfish in such a way that the paws press the claw arms tight against the side of the crayfish's body. The other eight legs are too feeble to offer any purchase or do any damage. Effectively the crayfish is strait-jacketed, the method working equally well whichever way up it is. All that is left to move is either the head or the tail. If it is the former, the little black pin-head eyes of the crayfish swivel out on their stalks, as if daring the otter to do its worst, whilst the two front antennae whip about in a last attempt to ward off the inevitable. If it is the latter, the tail waves around, but this is an utterly feeble defence. The position adopted, the otter is ready to eat the crayfish. Alive, of course. There is no attempt to kill it. If the unfortunate creature is head up, our otter starts with the head. If it is tail up, it starts with the tail. There is no sense of discrimination; it simply begins by chewing on the nearest body part and keeps going until it is all gone. The first few minutes are clearly a chore for the otter as the half-eaten crayfish, still in its death throes, tries to escape. But otters really do have this one down to a fine art. The crayfish stands no chance.

To make matters worse, if you have any feelings for crayfish (I know, it's hard …), otters are really quite measured eaters, making for a slow death. Vicious though their bites are, they tend to be small, tearing away at the victim's body then taking time to really chew up each mouthful. They have to do this because the gullet of an otter is small for an animal of its size; maybe half an inch in diameter, so the thickness of your little finger. Add to that the obvious difficulties of swallowing the shards of the crayfish shell (yes, they eat them whole), and you can see why they might take their time eating what might otherwise choke any other animal. It is a grisly thought that what goes in must go out, so otters cope with this by excreting a jelly-like substance through their digestive tract to, how shall we say, ease the passage.

Wisp's brother and sister soon joined her with the mysterious capture; they have a curiosity for anything new but none of them really had a clear idea what to do with it. They had it surrounded, occasionally darting at it without getting too close, whilst the crayfish showed a remarkable determination to defend itself with those pincers, successfully keeping the three at bay. Eventually Kuschta tired of their inability to work out the process, clambered down from her log eyrie, and gathered the crayfish to her chest to supply the gradual *coup de grace* as the pups looked on. Leaving the corpse almost intact, she wandered a little way off, settling down on the grass to groom and watch.

Now it is an interesting thing about otters that they are not overly aggressive about food. There seems to be a certain protocol that for the most part determines that whoever captures the food gets to eat it. Yes, sometimes one will attempt a steal, but a brief snarl will usually put an end to it. So it was that Lutran and Willow hung back. The crayfish belonged to Wisp who, imitating her mother, took a stab at eating it but soon abandoned the effort, the armour of the carapace shell too strong for her young jaws. Both of the others gave it a try, burying into the body cavity to seek out the softer flesh, but in the end the bulk of it fell to Kuschta, who finally devoured the entire thing.

There was a point around this time that the pups clearly frustrated Kuschta when it came to feeding time. She'd pull herself out of the river, fish in mouth, and deposit it with the pups. Sure enough, one would dig in, but as Kuschta returned to the river the other two would follow, with no apparent interest in eating. So she'd swim back to the bank and the two would do the same, at which point she'd unceremoniously grab each in turn by the scruff of the neck, fling them out of the water onto the grass where she would join them, forcibly nudging them towards the catch of the day. It seemed to me that their desire to be close to their mother trumped any hunger. And it wasn't just a nearness – they had to be touching close, jostling up against her the entire time. The same thing would be repeated again and again until Kuschta finally lost her patience, whipping round to drive them back to the food to prevent them following her into the water again.

But for all this apparent neediness the pups were learning some river craft amidst this play and effervescence. They had to; their future depended on it, and the early successes were largely at the expense of one rarely noticed but vitally important little fish – the bullhead. The bullhead *Cottus gobio* goes by other names such as the miller's thumb and freshwater sculpin. The miller's thumb probably best describes its appearance, based on the blackened, bulbous head and tapering body, the reasoning being that it looks not dissimilar to the thumb of a miller disfigured by years of being caught in the grinding machinery. But I like the bullhead name better, for it well describes the way it lives. It is an aggressive little bottom-dweller who likes the fast gravelly bed of the brook. It spends the day hiding out under stones, emerging at dusk to hunt for shrimps (yes, those again ...), nymphs, eggs and just about anything else that might be edible. They are not always easy to see, their blotchy brown skin merging with the stones, but once you spot one they are fascinating to watch. They bustle about the river, all muscle and spiky aggression, a bit like pumped-up nightclub bouncers. The big pectoral fins jutting out each side give them that ability to accelerate fast on their prey – they are ambush predators and take no prisoners. And when they are not on the look-out for food it is all about competition for territory; turf wars with other bullheads consume as much time and energy as food gathering.

The spawning season brings out an unexpected side to the bullhead; that of the concerned, expectant father. For most fish, egg laying is a fairly random endeavour – the two pair up, and the female sheds her eggs with the male at her side mingling his milt over them. That done, the parents have no further interest – the fertilised ova will tumble away to lodge somewhere on the river bed, where they may or may not hatch. But this is not good enough for our bullhead. He digs a shallow hollow in the gravel and invites a series of females to lay their eggs in his nest. Now this is not so unusual; trout and salmon do the same, using their tails to dislodge stones that cover the hollow when the laying is complete. But the bullhead goes one better, guarding the eggs night and day, warding off the many predators for whom eggs are a nutritious snack. The nurturing doesn't end there. To hatch, all fish eggs rely on a flow of well-oxygenated water, the life-giving oxygen percolating through the membrane. The better the flow, the higher the number of eggs that will successfully hatch, so our concerned parent takes this to heart, using his tail to fan the water over them, which he does in conjunction with his guard duties until they hatch two to four weeks later.

Maybe it is this extreme piscar parenting that explains the success of the lowly bullhead; healthy rivers are thick with them. In food terms they are what krill is to the ocean – food for the masses. Watch a grey heron patrolling the shallows and you'd most likely think he is after a big payday in the form of a good trout or something similar. The truth is that bullheads are populous, nutritious and easily caught. All the heron has to do is take position over a likely stone, move it to one side with its foot and then pincer the exposed bullhead with its beak before it has a chance to make off. Successful, the heron raises its pinkish-yellow bill to the sky, stretches its neck, swallows the bullhead whole and with a little congratulatory shake of the head takes a step forward to the next stone. A dozen bullheads are as good as a single small fish and a good deal easier to catch, so plenty of other water-fowl have a liking for our *gobio* too; aside from egrets they mostly aren't as good as herons at catching them, but the likes of moorhens, ducks and coots all seek them out, as do other fish. In fact, fellow fish make far more inroads into the bullhead population than the birds ever do; during some weeks of the year fish – trout, perch, grayling, roach – well, just about anything that swims, will feed on bullheads to the exclusion of all other food, given the chance.

It pretty well follows that what must be good for fish and birds must also be good for otters, though for Kuschta pick-ings would have to be pretty slim for bullheads to feature high on her hunting roster. But for the pups the novelty was impossible to ignore. At first the bullheads, forced to bolt by accident rather than design as the pups played in the gravel shallows, were largely a thing of curiosity, but that was never going to last for long. Lutran was the first to take to pursuit, jinking this way and that as a bullhead would scud away, tight to the gravel bottom, heading for the safety of deeper water. Very occasionally Lutran would win the race, but the net effect was mostly spray and foam, usually ending with him standing in the water rather mystified by the unsuccessful turn of events. To pay him his due he didn't give up at once; he'd remain rigid, poised for action like a pointer dog, three legs grounded and the fourth cocked ready to spring, staring intently at the last known sighting until he'd lose interest and wander off.

Otters definitely learn by a process: some instinct is clearly hardwired but it is not enough for a seamless transition to adulthood – that takes observation, trial and practice. So maybe it was seeing Lutran's lack of success that encouraged

Willow and Wisp to adopt a different, and very much more effective, tactic for catching bullheads. How they came upon it, I can't be sure. Maybe they watched the heron. The principle was much the same in that they crept up on a likely stone rather like a cat stalking a mouse. Then, with a flick of the paw, they would turn the stone over, leaping on the startled bullhead before it had a chance to flee by springing up then down to grab the unfortunate fish. As I say, I can't be sure how they came upon this hunting method, but my guess is that they'd been in the habit of turning over stones since they first ventured into the water, so the tactic simply evolved.

But whatever the evolution, soon Lutran was copying his sisters and for the next few weeks bullhead hunting was the game of choice. Rabbits, frogs, ducklings and plenty of others should beware – the pups were slowly learning the skills that were to eventually keep them alive – but as the family summer began to close out they had much, much more to achieve.

CHAPTER 8

LIFE IN THE FOOD CHAIN

It's rabbit for supper

It is an odd thing, but come September the valley seems to change overnight. One day every tree and plant is lustrous green, the next they have all taken on a very slight shade of brown. Leaves that were once bright and shiny, pliant to the touch, are now stiff. The fringe along the river edge no longer resembles that perfect herbaceous border. Most flowers are dead or dying, the stalks brittle and brown. Brush against them and a puff of dust will rise in the air for your trouble. We are still a little way off the first frost of autumn, but the evenings have a distinct chill. The evening dew is definitely cold to the touch, the night-time quiet, arriving fast as the bees and insects cease their hum just as soon as the warmth of the day disappears. The most active are the most silent, spiders venturing out to weave their

133

webs or tiny field mice gathering up an early harvest of seeds.

For the otters the dusk marked the end of another long day of doing very little, the habit of playing in the day now just a memory. In their respective resting places the four stretched and groomed, in no great rush to start out on the night ahead. The family had all but abandoned the holt for no other reason than that the summer made for better places. The Willow Island would always be there, but for now it was the sanctuary of last resort, the place to which they fled when dogs or humans came their way. Kuschta had made the willow pile her own. Some logs were sundecks, others shaded ledges. Deep inside was cool, the soft, rotting timber a comfortable bed with numerous ways in and out. The pups kept away. Lutran had found an almost ready-made couch, a long-disused swans' nest. Shaped like a small volcano, the two-foot-tall circular pile of reeds and twigs was woven tight, a yard or so from the edge of the stream and screened from view by bulrushes. The deep indentation in the middle was unsurprisingly more suited to cygnet rearing than a resting otter, so Lutran tore down the edges, packed the material into the hole, and levelling the top with dry leaves garnered from the rushes. Willow and Wisp showed less independence, preferring to stay close to each other, sharing the spaces within a long-fallen ash tree, hollow to the core, that lay along the bank just downstream from Lutran. The truth is that for all their apparent daytime separation the four were still very much together, resting within a stone's throw of each other, so come the first flitting bats Kuschta shook herself down, summoning the trio to her side with a brief, whickering whistle.

Heading off on their nightly travels, the otters are the

undoubted masters of this particular universe but they are far from being alone; the river at night is a busy place. Most residents of the valley will pause as they pass by, melting into the shadows – being unseen is the safe option. Some are competing for the same foods whilst others are trying not to be that food. As the otter troupe bundles through, it is clear that the valley is both shared and claimed. It is just a question of who dominates at that particular moment. If you respect your place in the riverine pecking order you might not be brave but you will survive. Otters are, like similar apex predators, well-adapted killing machines, but they don't do it for fun and rarely for territorial gain. They kill when they need to eat and they only need to eat when they are hungry. In a way there are similarities with the native pike *Esox lucius*, the alpha creature that rules the river beneath the surface. The brown trout might be the darlings of fly fishermen, the bullheads the essential diet for the masses, but it is the few solitary pike that keep a river healthy.

Esox doesn't have the best of reputations; fearsome teeth and a body that has barely evolved since prehistoric times are a PR nightmare. Nor does it help that pike truly do prey on cute, chirruping ducklings. Looking up at that silhouette formation line, the paddling little webbed feet breaking through the surface as they follow in the wake of the mother, must be all too tempting and all too easy for our predator. She lies in wait (nearly all monster pike are female; it is an evolutionary thing – males rarely exceed ten pounds), immobile for hours on end until the likely prospect comes into view. No pike worth its salt can ever resist, accelerating from stationary to killing speed in the blink of an eye, snaring a poor infant in a single snap of the jaws. And to make it worse, if you care for ducklings pike are patient, stealth killers who go about their work with little fuss; the brood will hardly know that one of their number has gone, so over a period of a few days a big pike will easily reduce a clutch of a dozen down to two or three.

Keeping a river healthy – it's a strange phrase and at first a hard one to put into the context of pike and their apex pals, who have the same duty in their respective ecosystems – lions in Africa, polar bears in the Arctic and so on. But pike do deserve better press; they have a job and they do it well. Those little ducklings are all very well but they breed like wildfire. If all twelve chicks survived to adulthood they would begat another twelve, and those twelve and another twelve, all within the space of a single summer. Without Mrs Pike taking her share, the population would explode with all the strains that would place on the available food and space. Pike do the same with fish, but with a little twist, picking off the diseased and less able, ensuring the local population gets stronger over time through that Darwinian imperative. And as for the pike themselves, well, they are not without some sense of their own mortality. When times get tough they will happily eat their own to weed out the competition.

So what is it that otters do to earn their place at the top of the food chain? Well, it's probably easier to explain it through the tale of their evolutionary cousins, the mink. Mink are not native to the British Isles and the ones you'll see are American mink *Neovison vison* which are different to European mink *Mustela lutreola* who have never, despite their name, populated Britain. Mink have a lot in common with otters: they too are semi-aquatic carnivores preying on fish, crayfish, birds and small mammals. They like life by a river, calling the valley home and rarely straying far from the scent of water. They mark out their homelands with scat, and the males are fiercely territorial, permitting just appropriate females within certain areas. When it comes to breeding they are less robust; litters are only born in the early summer, but like certain types of otter they have that ability for delayed implantation. So far so familiar. But in size they are an awful lot smaller, no more than a fifth of the weight of otters, though still two feet in length. It is probably the long body and dark brown fur that, when only given a fleeting glance, makes it easy to mistake them for otters. If you are in any doubt, try to grab a good look at the tail, it being more catlike than otter. But if you see them clearly (they get out in the daytime far more than otters), aside from the size, the body shape is very different to that of an otter. It is often said they are somewhere between a cat and a ferret; I'd err more to the ferret side of the equation, with a slim body but without that humped, strong rear carriage. If, like the pike, it gets bad press for being vicious, it certainly redresses the balance with an appealing, inquisitive face with fluffy half-moon ears, bright black eyes, a pert wet nose and long whiskers that would melt most hearts. And it does, of course, have the most gorgeous fur of any animal that walks the British countryside.

It is often assumed that mink in the wild is a relatively new thing in the British Isles; in the 1990s thousands escaped from breeding farms at the height of the anti-fur trade activism when protesters broke in, breaching fences and cages. It caused all sorts of outcry at the time. The mink, unable to survive in the wild, mostly died, but not without wreaking havoc in the immediate neighbourhoods. So the protesters, having thought they had done a good thing, were lambasted on three fronts: taking the law into their own hands, the cruelty of releasing animals unsuited to life in the wild, and the damage to other wildlife – mink have a penchant for birds' eggs, for example.

But whatever the rights and wrongs, the tide was running fast against fur farming, which was made illegal by an Act of Parliament in 2000.

In fact, mink were first imported way back at the start of the twentieth century, and the first farms were established in the 1920s. As an industry it took off; mink are easy to breed and handle, and are ideally suited to the British climate. Though they mostly eat meat or fish in the wild, in captivity they thrived, rather bizarrely, on cheese and dairy waste products. After the hiatus of the Second World War the number of farms continued to rise, peaking at around 700, along with numerous illegal farms. Whether it was the illicit farms, or just the inevitability of it, mink soon escaped and a self-sustaining wild population became part of the countryside. It seems likely this had happened within a decade of the first farm opening, though it took until 1956 for this to be officially accepted, and a survey in 1967 concluded that mink were present in half the English counties and most of lowland Scotland.

If these dates seem familiar it is because the decline of the otter population almost exactly correlates with the rise in the wild mink population. Connection? Well, there almost certainly is. A river valley devoid of otters was the perfect habitat for the mink, who moved into the vacant space. They soon became an accepted, if not loved, part of the rural fabric – so much so, in fact, that the otter hunts turned their hounds to pursuing mink. One pack even went a full decade chasing the newcomers without sighting a single otter, such was the dramatic change in the fortunes of the respective mustelids. The mink proved to be highly adaptable; the escapees from the very early days of mink farming provided the blood line for the wild population that remains today. American mink are naturally dark brown, a chocolate colour most people would say, the fancy furs of sable, white, blues and so on being the product of selective breeding. But clearly such colours are not well adapted to life along a lush river bank; the greatest threats to the mink are hawks and owls who are able to pick them off all too easily. So, self-preservation being what it is, within two generations the mink had reverted back to their natural brownish hue.

Though not an apex predator, the mink do have some advantages over otters; they live longer to start with, somewhere around eight to ten years compared to four or five. The litters are larger, the young reaching maturity and independence in months rather than years. They don't need the same quantity of food, so bad times are easier to survive. And with certain food sources – water voles, waterfowl and eggs – they have a distinct competitive advantage over otters. When it comes to raiding nests they are better than otters at climbing trees; come to think about it, I have never seen an otter do much more than clamber along a fallen tree trunk, let alone scale any great heights, though I'm sure it is not beyond them if they have to. Mink, smaller and less conspicuous, are much better at hunting mallards, moorhens and coots, all of which are a main part of their diet, as are water voles. The latter are the best illustration of how pivotal otters are at the top of the food chain.

There is definitely a pecking order to all the creatures sighted in the valley: otters are the most impressive; physically imposing, the rarity and fleeting nature of sightings adds to their mystery. Kingfishers, an exocet of blue light, are annoyingly transient, always leaving you wanting more. Ducks, moorhens and swans all have their moments – mostly when chicks and cygnets are around – but they are everyday inhabitants, although none the worse for that. Owls and hawks catch the eye. That hovering flight is impressive, but it is the purpose that draws you in. Death. The aerial hunter poised fifty, sixty feet in the air, static bar some subtle wing adjustments for the perfect hover on the wind, whilst picking out some tiny scurrying mammal that thinks it is safe amidst the grasses of the meadow. Then quite suddenly, without any apparent warning (at least to my eye), the wings collapse, the bird plunging towards the target. With impressive precision, never pausing to make contact with the ground, it plucks up the prey, squeezing the life out of it in the grip of its talons as it flies off to some handy roost.

Amidst all this lives the European water vole, *Arvicola amphibius*, or Ratty of *Wind in the Willows* fame. He really is the cutest of all the valley creatures, with bushy, chestnut brown fur covering a chubby little body, black dots for eyes looking out from a whiskery face, with a tiny nose and furry ears. They are buoyant things when they swim, their stretched-out bodies arching slightly, three-quarters out of the water, the four legs furiously paddling them along, nose held aloft, a tapering tail behind. They are intrepid swimmers, crossing wide rivers and lakes, but they generally prefer to hug the edges, weaving amidst the vegetation along the bank. All in all, they live fairly blameless lives, vegetarians through and through. Grasses, roots, seeds, wild fruits – well, just about anything that comes their way – at the last count they regularly ate over 200 different items of food, so they don't do too badly despite a diminutive stature. There is a slightly human aspect to the way they eat. They sit back on their haunches, using their front paws like two little hands, gripping a reed stalk or maybe a blackberry in autumn, which they chew at with rapid little bites, emitting a sort of wet, smacking-of-lips noise as they go. They are not very quiet about this – on a still morning you'll hear them long before you ever see them.

Of course, these voles are not entirely benign. They live underground, burrowing into the soft earth of the river bank to create a labyrinth of tunnels. Similar to an otter, they try to create an underwater entrance, so don't be surprised if, should you scare one, it plops beneath the surface never to reappear. They breed fast, though – two to five litters a year with eight offspring each time being normal – so numbers can multiply, creating huge colonies, the tunnels collapsing banks during times of flood. However, mammalian earth-works are the least of our worries today because, when it comes to water voles, mink have cut a swathe through the population, putting it on the endangered list. The numbers continue to remain low today, having declined precipitously by 90 per cent in the 1980s and 90s.

For mink, water voles are high on the food list, and the fact that mink moved into the space vacated by the otters does not bode well for these aquatic creatures. Certainly otters do eat water voles from time to time, but mostly they rub along just fine – not least because one occupies the day and the other the night. The only time I saw any voles in any danger was when the pups were first finding their way, playing during the day in The Badlands. That said, you'd be a pretty unlucky vole to stumble unwittingly into the path of that boisterous trio, though Lutran did chase a couple down for the hell of it. However, sadly for our little furry friends, the mink is a well-adapted hunter with two abilities that set it aside from most other local predators: it can swim and it can squeeze into the vole burrows. Otters can do the former but not the latter. In a matter of a few days a single mink is easily able to wipe out an entire vole colony, searching the tunnels for the nests, eating an entire litter at a single sitting. To make matters worse, unlike otters, mink kill for fun, so even when sated the destruction continues.

To this Armageddon you'd think water voles have absolutely no defence, and in the specific sense that is true, but the revival of the otter population has halted the mink in their tracks. There is no way, despite or perhaps because of their common weasel lineage, that the two species are able to occupy the same territory. The otters left and the mink arrived; now the otters have returned and the mink are leaving. Otters, considerably bigger and stronger, are perfectly happy to go head to head with mink, and there is no doubt that otters regularly kill mink. The natural order is gradually returning, the otter proving its apex worth. Ratty and his friends are coming back, as are the moorhens and coots, as the egg-stealing, baby-eating mink are marginalised.

Of course, otters are not always guardian angels, as the rabbit population will attest. At first glance, rabbits might seem an unlikely meal for otters; in habitat terms they are odd bedfellows and, to be quite frank, who would bet on the otter in a straight race with a rabbit? But in the Wallop Valley they share much of the same territory, and so it is hard to see that such a tasty meal would go unexploited. But how and where? Well, if you took a slice of the valley it would look much like this as the land radiated out from the river: water, river bank, woodland, headland, arable field, headland again and then hedge. In my experience, we rarely see rabbits close to the brook, though you'd think the thriving vegetation would be a tempting prospect. My assumption would be that, as animals of flight, the barrier of the river constitutes too much of a danger if attacked. Yes, rabbits can swim, but I suspect it would be a last desperate act rather than a default choice. The woodland, with the exception of the edges, is mostly avoided. There are too many opportunities for stealth hunters like foxes and stoats, the straight-line speed advantage of the rabbit negated by the terrain.

No, where the rabbits really like to dig their burrows is at the edge of the woods or in the hedges, with just the headland between them and open fields. Headlands are one of those relatively new aspects of the farming landscape that you would not have seen back in the 1970s. Up until that point, in the drive for maximum yields and before the concept of biodiversity had an audience, farmers ploughed, planted and spayed every square foot of soil. Today it is a bit different. If you look at a ploughed field you will see a strip of land around the edge, anywhere from five to twenty-five yards wide, that is different to the bulk of the cultivated land as it has, in some way or other, been left to nature. It takes all sorts of forms; sometimes it is a strip of grass, other times wild flowers, less often bare earth or at other times a mixture of the crop and wild plants where the strip has been left unsprayed. Collectively this has been a huge boon to wildlife, from the very smallest insect to those right up at the top of the food chain, rabbits included.

That said, they have not reciprocated the favour to farmers by leaving their crops alone. Far from it. The headlands, especially the grass, provide a handy staging point between home and food. An open strip of land is safe; predators are easy to spot, with a short dash to the burrow always being the flight option. Next time you are near a field in spring, take a look at the crop just alongside the headland. Compared to the rest, it will be stunted at best, non-existent at worst.

Rabbits make short work of the freshly growing shoots of wheat, peas, barley or whatever crop it might be, keeping them constantly nibbled down. And they rarely recover. If you return close to harvest time, the beautiful sea of swaying yellow corn will be framed by a very ragged edge of stunted stalks all around the headland. Farmers pay a price for feeding our growing rabbit population.

But they are doing Kuschta, Lutran, Willow and Wisp a favour. There is still some debate as to exactly how much food otters eat daily, varying from 10 to 20 per cent of their body weight. In truth, nobody really knows, because the data garnered from captive otters in no way reflects the lifestyle of their wild counterparts, who are very hard to monitor accurately. But let us make some assumptions. By late summer the Wallop four will be weighing in collectively at somewhere between thirty and thirty-five pounds. Kuschta will be half that, with Lutran around seven pounds and the sisters each a pound or so behind him. Using all that, plus the fact that the pups are growing fast and expending a huge amount of energy enjoying life, and it seems fair to assume that at the very minimum they require four pounds of food daily to survive. But they really need to do better than survive, so they'd most likely be at the upper end of the scale as they put on weight for the privations of the winter ahead.

In terms of the lake at the mill, four pounds is a slam dunk – two fish and you are done. But strangely, however easy the pickings, after that initial burst when Kuschta first introduced the pups to the lake, their visits became less frequent; maybe one night in three was the norm for now. Maybe it was too easy, not enough hunting craft learnt, or perhaps it's that 'avoidance of over-exploitation' gene – an otter version of not killing the goose that lays the golden eggs. Whatever the reason, they were clearly finding enough food elsewhere, but was it from the Wallop Brook? Four pounds, maybe rising to eight, night after night, is a big call. First of all, the trout are small and wild; any fly fisherman would mark one of the beautifully speckled three-quarter-pounders down as a catch of the season. So, let us say half a pound is the average. On that basis, and assuming the otters could catch them (questionable, as small wild fish don't have much trouble outwitting otters), the river would be stripped of edible-sized fish within a month. Eels are an option, but again a good-sized eel is only half a pound and by September the run to sea is starting to ebb. Of course there are all the other things – small mammals, birds, crustaceans and so on – but they are not necessarily the high-protein diet an otter needs. Even if they went on a crayfish binge they would have to eat somewhere in the region of 300 a week, playing hell with even a digestive system as robust as an otter's.

As none of this is going to happen, at least for any sustained period of time, it is pretty clear that otters look to places other than the river for food. So, amongst more earthbound prey, what could be more nutritious than a rabbit? An adult weighs in at anywhere between three and four pounds, close to the whole day's ration for the four in a single hit. It might seem an unlikely target, but it is the rabbits' misfortune to be in that group of animals classified as crepuscular. That is to say, active in the time immediately before dawn and after dusk, more than perfectly dovetailing with otters, who like nothing more than a burst of activity at twilight, a rest during the dead of night, with a last meal at dawn before settling down for a day on the couch.

Rabbits are not much found in the water meadows or the places otters really like to live; you'd never see one in The

Badlands, for instance, so they were something of an unknown to the pups. There were some around in the Beech Wood hollow, but that was in a time long before they started to hunt for food. Their first encounters were accidental. Those times when they strayed from the river corridor, exploring beyond the narrow strips of woodland, popping out the other side into the fields. Quite what an otter thinks of a rabbit on first sight is hard to know, but Willow clearly thought them nothing odd, padding along the grass strip towards them, more curious than anything else. The rabbits, on the other hand, knew better, pausing their grazing to observe the oncomer. One of them sits up on its hind legs, better to see who is coming, ears pricked and twitching for alien sounds, front paws held up, poised to flee if the indications are bad. Willow never really stands much of a chance. Rabbits don't have a great reputation as higher-order animals, but that is not entirely fair as they are actually well adapted to the life they live. With close to 360-degree vision, with eyes on top and to the sides of the head, they can at the same time detect predators (they are far-sighted as well) whilst working out an escape route. So, whatever the verdict the lookout came to, even though an otter must be somewhat unusual, it is not exactly mass panic. The lead rabbit drops down onto all fours, showing the way as they all bounce across the turf to the burrows, disappearing long before Willow gets close.

Willow seems nonplussed at their sudden disappearance. In the world of the otter this makes little sense. If you don't get your prey first time, you at least get a chance to chase it. Sniffing the various burrow entrances was clearly no more enlightening, so after a while Willow got bored, heading back through the woods to the river. But otters are nothing if not curious and persistent, so a few dusks later she was back again, this time decisive, accelerating to a run the moment she spotted the rabbits. The result was exactly as you might imagine. Otters can gallop reasonably fast, maybe 10 – 15 miles an hour for short bursts, but they don't exactly cover the ground with great quiet, the vibrations of her pounding gait alerting the rabbits to the danger in a trice, and they vanished underground. However, Willow was not about to give up that quickly – stealth had worked on bullheads and frogs so why not rabbits? Why not indeed. Well, you couldn't fault Willow's attempts at stealth, her catlike approach through the cover of long grass bringing her closer to the rabbits than ever before. Her Achilles heel was not her tactics but just the way otters are; they have a damp, musky odour. It might not be unpleasant, but it is certainly unusual and rabbits don't twitch their noses for no good reason. They have highly advanced olfactory senses, ten times better than humans, with smell being as much an indication of danger as sight or sound. Round three to the rabbits.

Naturally enough, it was Kuschta who showed Willow, and by default the other two, how to catch rabbits. The answer is blinding in its simplicity. As she led the four along the track there was no attempt at concealment. No running. No commando-style approach. The rabbits fled, as they would, and that, in the normal order of things, should have been that. In no great hurry, Kuschta appraised the various warren entrances, sniffing at each, trying them for size with her head and shoulders. The pups got in on the action, coming at her from the sides, scrabbling to push into the entrance at the same time, plugging the hole, until Kuschta backed out, leaving the pups in a tumbling heap as she moved to the next burrow. Sometimes one of them might linger alone, inching a little further into the darkness, but without mother at their side the adventure was soon abandoned as too dangerous, the pup running to catch up with the others. The inspection complete, Kuschta returned to one particular burrow, and as if sensing there was something different this time, the pups obediently held back as she disappeared head first into the ground, vanishing entirely from view.

Goodness knows what chaos ensued down in the burrow. Kuschta's advance must have been signalled on every level: noise, smell and eventually sight. Flight was always an option. Warrens tend to have multiple ways in and out, but equally they have many cul-de-sacs. Kuschta would easily fill the tunnel. There would be no way past. Cornered, a rabbit can lash out, but their paws are designed for digging, not fighting. In any case there is barely room for much combat. Kuschta can soon subdue even the biggest rabbit with her ten sharp claws, pressing it down until she can get to work on the vulnerable head or neck with those powerful jaws and four vicious incisors. Rabbits rarely go quietly; their death scream is akin to a wailing infant, their hind legs scrabbling for purchase as Kuschta alters her bite so as to eventually sever the spinal cord. She knows she has every advantage on her side whilst in the confines of the burrow, so it is not until the rabbit is entirely dead that she drags it to the surface.

The first taste of rabbit must have been entirely unlike anything Willow, Wisp and Lutran had ever eaten. Sure, they would have had a few small mammals – field mice, water voles and their like – but I doubt anything could have prepared them for the warm-blooded carcass of a rabbit. But, unfamiliar or not, all three tore into it, biting, chewing, grinding and licking, little tufts of grey bunny hair drifting in the air until the only bits left were the leathery ears, white tail, some small part of the intestine that I can't recognise (and they never eat) and a bloody mess on the grass. From that day onwards rabbits were to become a staple of their diet. At first Kuschta did all the catching, but soon the pups took to the burrows. Lutran, bigger, stronger and generally braver, was the first to strike blood, snaring a six-week-old that was unfortunate enough to flee in the wrong direction. It wasn't always a win for the otters; adult rabbits could generally put up enough resistance to ward off Willow and Wisp in the warren, or push past to escape. But where they lost out on the big game the pups gained with the kittens, using the advantage of their smaller size to reach the baby rabbits in the deepest recesses of the burrows.

The downland of southern England is a stronghold for rabbits; the landscape is particularly to their liking and it might be tempting to think that, with lives in such close proximity to otters, becoming a source of food is strictly a localised thing. But it isn't. Rabbits are standard otter fare across the British Isles, even in parts as northerly as the Shetland Isles, where one radio-tagged adult female otter caused a certain amount of consternation with researchers, disappearing off the radar for a little while each day in more or less the same place. A little good old-fashioned human observation revealed the reason; the otter was making a regular visit to the local warrens, where his tracker stopped working when he went underground.

Of course, our rabbits have not always been so populous. Actually they are not 'our' rabbits at all. The European rabbit, *Oryctolagus cuniculus,* is not native to Britain, originally imported from the continent by the Romans in the first century or the Normans in the twelfth, depending on whom you believe. True to their reputation (one female can easily give birth to 24 kittens in a year), they soon took hold, establishing their place in the countryside with no great difficulty until it all went wrong in the 1950s when the myxomatosis plague struck. Myxy, as it is often known, is a particularly virulent disease that affects just rabbits, and it is unpleasant in every respect. I remember it well growing up. Blinded rabbits, immobile by roads and hedges, their eyes puffed closed with puss that ran down their faces like tears. Their fur starry and dishevelled, the worst cases covered in fleas and ticks. From myxy there is no recovery; death is slow (they essentially starve because they can't see to eat) and inevitable, taking anywhere between a few days and two weeks.

Like the rabbits themselves, the disease was imported, arriving in 1953 not long after it had been deliberately introduced to Australia as a form of government-sponsored pathogenic control. Nobody quite knows how it came to Britain (it is illegal deliberately to release diseased animals), but we do know that a retired French bacteriologist injected two rabbits with the virus in the summer of 1952 to control the rabbit population on his estate 75 miles south-west of Paris. Within six weeks 99 per cent of them were dead, but his plan went awry as the disease swept across Europe. Ironically the bacteriologist, Dr Paul-Félix Armand-Delille, was both feted and vilified in his own country, condemned by rabbit hunters and praised by farmers. In 1955 he was convicted and fined the not inconsiderable sum of 5,000 francs. The following year the Agriculture Ministry awarded him with a gold medal, striking a special edition with his head on one side and a dead rabbit on the other. *Vive la différence!*

However it arrived on British shores, myxomatosis was first discovered in Kent and it proved to be devastating. It is a disease that is spread by contact. Life in the close confines of warrens is the perfect incubator. Within two years 95 per cent of all rabbits in the British Isles had died, and there was serious discussion as to whether they would become extinct.

Why, you might justifiably ask, is this relevant to otters? Well, with horrible timing, just when the otters were being hit by the pesticides, one of their single most important sources of food was being wiped from the face of the countryside. It was a perfect storm and, as the saying goes, if it wasn't for bad luck, they wouldn't have had any luck at all. Frankly, it is amazing that otters have survived, let alone clawed their way back to the point where they might almost be described today as thriving. But if the otter faced a struggle to survive, so did the rabbit. Through the next two decades their numbers bumped along the bottom, but gradually that rapid turnover of the generations came to their aid as they became immune to the virus. Today the population is back to pre-myxomatosis levels, and though that might be cursed by farmers and gardeners, there can be little doubt that their revival has helped the otter.

CHAPTER 9

TO AUTUMN

Equinox

When Keats walked the banks of the River Itchen, near Winchester, close on two hundred years ago, composing his famous poem 'To Autumn', he was just the other side of the hill and a handful of miles from the Wallop Brook. In those three stanzas that open with the line 'Season of mists and mellow fruitfulness', he captures a countryside that is, in many ways, not so very different today, the progression from the warmth of summer to the impending winter as true today as it was then. It might be tempting to write that the trappings of human progress have taken a fatal toll on his words, but they haven't. The cottages are still thatched. The poppies still bloom. The full-grown lambs still bleat. Of course, for many of us, more removed from everyday rural life than we might like, the subtle transitions from summer to autumn and then winter are not always easy to discern. But we do notice the changes when writ large; the first chill frosts, hedgerows laden with fruits and berries, the gust of wind that strips trees of a few early browning leaves that flutter to the ground, the shortening days and lengthening nights.

But for the creatures of the valley the countdown from the time of plenty to the depravations of a bleak winter is marked with daily changes. How they adapt and, in that way that animals do, plan, will determine if they survive to the following spring.

For the otter quartet those choices were evolving, though in different ways for each of them. The burden of hunting still fell on Kuschta. True, she could now travel with her attendants in tow – no need to lug a fish corpse a mile or more – but in the final analysis, despite the pups approaching their six-month birthday, she was still the bread winner. Lutran was becoming increasingly rumbustious, differentiating himself from the others by his behaviour. It would be extreme to describe him as a bully or aggressive towards the others (that will come in time), but he was unruly, disruptive, throwing his weight around for the best space in the holt for no other reason than because he could. His attempts at hunting were often furious rather than effective, making life harder for his siblings as he spooked potential prey, so on the night-time travels they began to split into two tribes – Lutran and the rest. Willow and Wisp, by contrast, still shadowed their mother, content to follow her lead. In appearance they were losing their puppy fat, becoming more like complete half-size adults, whilst Lutran was long in body and skinny in frame with plenty of filling out to do. But that was fine; there was still plenty to eat for now.

I don't know it for an absolute certainty but I reckon it is a fair assumption that September into October must be one of the most bountiful periods of the year for everyone and everything that lives in the valley. It is the time of year when you will see more fish in the river, birds in the trees and animals on the ground than at any other time. As Keats would say, all very fruitful. And that's before we even think about the tiny things that live in, on and under the ground or water who are in profusion themselves. It is, of course, a preservation mechanism – a huge overproduction of offspring in the hope that a few will survive to the following spring. It is a fairly brutal attrition rate even in the relatively benign climate of southern England where, for instance, only one in five blackbirds will make it through the winter.

But they all get a good shot at making a go of it; there is more than enough food to go around for now, the chance to build up the body reserves to be in the best condition when the worst comes around. The herons and egrets stalk the shallows from dawn to dusk, acting as if all their Christmases have come at once as every ditch and stream is packed with delight – frogs, newts and numerous small fish – all there for the taking. The kingfisher is beside himself with excitement, flashing from perch to perch, tracking the shoals to take his fill with ease. Even the moorhens, confirmed vegetarians, don't go short, filling up on the seasonal seeds and pods. Along the banks the water voles join in, seeking out the fruits of autumn – blackberries and hazelnuts are favourites – to give a timely boost to their usual drab diet of grasses and roots.

In the river it is a similar story as the fry of the spring grow into fingerlings, feasting on a stream packed with bugs and nymphs that have also done well over the summer. Of course, it is not all good news for the young fish – at a certain size they become prey for larger fish and the predators such as the aforementioned kingfishers, but such is the nature of life in the food chain. Out in the meadows it is something of a similar tale; the field mice, like the voles, are making the most of the bounty, but they have to take care as the barn owl chicks of April are now the young adults of September, with all the consequent hazards that poses.

The climate is just turning down a notch as the long, lazy days of summer become a slowly fading memory, but against that the weather is without extremes. Across the valley the fields are shorn of their corn. Some have already been ploughed, sown with next year's crop awaiting a good downpour that will breathe enough life into the seed for a burst of growth before lying dormant over winter. Other fields are covered with raggedy stubble, delineated by the track marks of the combine harvester, its work done for another year. Little bursts of colour dot up between the stalks; a few rogue oilseed rape plants, in a last-gasp attempt to grow and seed without the competition from the tall wheat, burst out bright yellow. Some stunted red poppies do the same, as do fast-growing weeds like thistles, docks and nettles. Huge stacks of straw bales, bigger than houses, loom up here and there. Some look alarmingly unsafe, the sides way off perpendicular, but nobody worries much. They will be gone soon enough; food and bedding for cattle in the months ahead.

The grass meadows that roll down to the river look neat and tidy, which is something of a shock. For months you have become used to the untidy patchwork left by grazing livestock – the tall clumps of nettles they choose to eat around, the scrappy stalks of ragwort, the dark green patches where the grass has overdosed on the cow dung fertiliser. All that is gone now, topped level by the grass cutter that has scattered in its wake shards of dried cuttings that will soon be gathered up by the field mice and their like as they build nests for winter. And all this enveloped by the smell of new-mown grass which plays a trick on the mind, provoking memories of summer at a time of autumn.

The trees are still undecided about the changing of the season. Some, like the horse chestnut, are harbingers of what is to come, the leaves turning brown and brittle long before it is the decent thing to do. But for most the change is not that pronounced, though you can tell at a glance that the sap is ebbing rather than rising. A month ago, in the early morning light the leaves sparkled, the dew capturing and reflecting back the first sun rays of dawn. Now that fresh lustre is gone. Certainly the leaves are still mostly green, now a shade on the side of camouflage. It's a little sad to see the future foretold in the branches, but once a few hard frosts have turned the woodland from drab green to that gorgeous golden of autumn the melancholy will be forgotten.

Below the trees the Wallop Brook is at its lowest ebb. Nature's water pumps, those aquifers that spring from beneath the chalk downs, are beginning to run on empty. The months without rain have taken a toll, and salvation – the rains of winter – is still a little way off. But chalkstream are resilient; they have a way of coping with these seasonal changes that preserve the life of the river and the creatures that depend upon it. Back in the spring the brook was wide and full, the banks were bare and the water deep, pushing through with vigour. The force of the water keeps the marginal plants that grow along the bank at bay, for they literally can't get a foothold; tentative tendrils and roots are washed away. But in time the flood force of winter abates, the flow ebbing back a few fractions of a knot each week. Small changes make for big differences; slowly those plants that appreciate life half in and half out of the water start to take a hold. Flag irises are the first, the green spear-shaped reeds pushing up from the river bed, emerging from the rhizomes that have spent the winter dormant in the mud. They grow so fast that they seem almost to rise up before your very eyes until suddenly the bright yellow flowers bloom, a burst of colour along the otherwise uniformly green and brown river line of April.

These standard-bearers are not alone for long. With each passing day the pace of the water eases. Slowly the character of the brook changes. No longer is there a harsh delineation between river and bank; the two are now connected by a ribbon of low-growing plants that are only partially rooted to the river bed, floating like a mattress between land and water. Step on them and they will give you no support, your foot crushing the leaves against the gravel bed. Don't be surprised if unfamiliar smells float your way as you go; the menthol of water-mint, *Mentha aquatica*, is as pungent as any peppermint tea. The peppery scent from the watercress, *Rorippa nasturtium-aquaticum*, will irritate sensitive noses, and you may see an oily slick curl out on the water. This spreading light-brown haze is exactly what you might imagine it to be – watercress oil. Apparently it has all sorts of amazing herbal properties for weight loss, as an aphrodisiac and as a baldness remedy. I can't attend to the veracity of any of these claims and nor do the river creatures seem to know anything different to me – I can't ever recall seeing a bird or animal tucking into a meal of cress. That said, the plants are magnets for nymphs, shrimps and all kinds of bugs; the moorhens spend their days dipping heads between the leaves for food, and that layer of water between the gravel and the plants is a haven for young fish, largely sheltered from predators and with an abundance of easily found food. In the late summer the mint will flower, poking up a tall stem from which sprouts a head of crimson-red buds that within a week burst like a firework into a ball of white-mauve flowers that will soon be covered with ants and bees in search of spearmint-flavoured nectar.

Not all the plants excite the nose or taste buds. Some are just simply lovely. Water forget-me-not, *Myosotis scorpioides*, is very similar to the land plants of the same name that spill over the edge of flower beds in traditional cottage gardens, serving a similar purpose as the pretty frame around the bigger picture. Like their garden cousins, the water plant grows in profusion. Horticultural books describe their spread as 'huge' and they will keep growing until they meet a greater plant force – they are far from being invasive. That wouldn't be the nature of forget-me-nots – they are too delicate to bully others, the clusters of sky-blue flowers with a yellow centre contrasting with the bright green leaves that are shaped like a lamb's tongue, a favoured egg-laying spot for newts. Blue water-speedwell, *Veronica anagallis-aquatica*, is the fourth component in this spreading blanket, but it is the scruffy one. Certainly it has white-blue flowers, but they are insignificant against the bulk of green, spindly, leafy stems that grow up then fall over or creep out sideways.

But beautiful or not, the growth acts on the river, narrowing it to constrict the flow. Look up the same river in mid-June and the width of the silver line of water down the middle will be half that of April. Fast forward to September and in the really shallow parts the watercress in particular will have grown almost all the way across, just a narrow meander of a stream left to connect the deeper pools. This might look bad but it really isn't. The vegetation is preserving the depth, holding back the water like so many natural weirs. The plants are creating shade, keeping the water cool, especially through the burning sun of August. And even though it might not look like a river, it still is. The water, though often hidden from our view, is still flowing. Fish are moving up and down, the plants providing a handy cover against their enemies who would otherwise scoop them up in such shallow water. And of course all the other tiny aquatic creatures are thriving in this watery jungle.

But, as ever with nature, it doesn't last. The first heavy frosts of the year will kill the watercress because this plant is, no surprise, predominantly made up of water, and one good freeze turns the leaves limp then rotten, the stems following suit. Before long the winter rains will come, washing away the debris, soon making the watercress just a memory. The mint, speedwell and forget-me-not will hang on a bit longer; they are more resistant to the cold, having put down roots and gathered beneath them a silt bed on which to grow. But eventually the force of the water gets too much; the silt is washed away, the roots exposed and one by one the plants get carried away like tumbleweed blown across a desert. Three months on, the brook will have gone full circle. Look up the stream at Christmas and it will be full, wide and fast again. The banks will be bare, the plants that did so much are no more, for now.

All that is in the future, though, and right now it is simply a race to prepare for that eventuality, even if not everyone feels quite the same urgency. Mion, the absent father of the pups, was content to take it easy – with at least four winters behind him, the change in the season held no great fears for him. Since mating with Kuschta, he'd been around the mill for sure, but he was harder to track because he arrived alone, without the chattering accompaniment. That said, he is not the quietest otter on the planet; coughing and wheezing as he hunts, more akin to a bronchial old man. Whereas the others came and went, he was more inclined to stay, the mill wheel house his particular favoured spot. He was also more inclined to head further up the brook towards its source, happy to take his chances through our village and the next, Over Wallop, something the others rarely did as far as I could tell. I have a good idea why he did this; whereas the others shied away from travelling past the homes that crowd in close to the bank and the road that runs alongside in parts, for Mion, travelling solo, the lure of another trout pond (it is tiny; I've seen larger swimming pools) was worth the evasions required. It is a hobby lake, a bit of fun for one of the local farmers. Well, fat trout in close confinement, it doesn't take a genius to work out the appeal for Mion. In an effort to keep him out, the farmer has surrounded it with electric fencing, the sort you'd often see around a field, but in this case two strands of plain wire four and ten inches above the ground. Does it keep him out? Well, no. But sometimes it just makes you feel better to know you have tried.

That is the thing about otters. They are persistent when it comes to food, not least because they require so much of it. At trout farms it is a war of attrition, often with many acres of ponds to preserve. Not only do you need tough fencing – an otter can easily scale a four-foot-high chain-link fence or tunnel beneath it – but every inflow and outflow is a potential weak spot that the likes of Mion will exploit relentlessly. And when you consider that your average trout farm requires the throughput of millions of gallons of water a day, all straight from the river, you can see how hard the defences are to plug. Steel culverts with bolted-down grilles are the order of the day. Faced with this, most trout farmers take a pragmatic approach, creating a 'sacrifice pond' on the otter side of the fence. One of the things about trout farming is that you have casualties; deformed fish, poor growers, runts that are attacked by the others, minor diseases and that sort of thing. In most respects these fish are perfectly healthy, but not of a standard fit for the table or stocking. So rather than knocking them on the head, these are consigned to the sacrifice pond, as easy pickings for passing otters. As far as they are concerned, a trout with a gammy tail tastes as good as a fin-perfect one (and is also probably easier to catch ...), so if it all goes according to plan, they won't bother with all the effort of breaching the defences of the farm to reach the healthy stock. As the song says, a few must die so the many can live.

That said, I often wonder if there is a vindictive streak in otters. Dave, one of my near neighbours, has, or I should really say had, a small pond in his garden that ran down to the brook. In it were a dozen or so koi carp, none huge, maybe the longest at twelve inches, that had been treated like pets for years, living a blameless and unmolested life. Then one or two disappeared. The heron was fingered as the culprit, so protective netting was draped across. Nothing too industrial, just the sort you'd use over fruit bushes. A few nights later this was torn away, like a blanket stripped from the bed, and a few more fish disappeared. Dave at this point knew he was up against more than just the heron; the noise of the eeking raiders had reached his ears as he lay in bed during the dead of night. So, heavy steel mesh weighted down with good-sized bricks was deployed, Dave confident that the remaining koi would be safe. But they weren't. Over three nights the otters worried, pushed and strained against the defences for hours on end until finally they found a way in, cleaning out the last of the fish. Now unless there is something very special about koi that I don't know about, there seems to be little reason why the otters became so determined. The mill lake is no more than fifty yards from the koi pond, completely undefended, where one of the bigger trout would weigh as much as six koi put together. Maybe there is a smell to koi they can't resist? Or maybe the otters presumed the pond held more than it did? Perhaps it is an apex predator thing never to be thwarted? Whatever the reason, one way or another they were not going to let the koi go unexploited, and so Dave, accepting the inevitable, has filled his pond in for lawn.

Mion's summer was one of a grand emperor touring his empire. The range of his travels is really quite remarkable. There are over twenty miles between the farthest extremes of his territory and within a week he'll visit those and all points within. So, in terms of what I know, that is the entire Wallop Brook plus some miles of the much bigger River Test, both upstream and downstream of where it is joined by the brook. It is not easy to be exactly sure how many 'families' Mion has under his purview at any one time. As far as the Wallop Valley is concerned, I'm reasonably certain it is just the one: Kuschta and the pups. I know they may sometimes disappear from my view for a week or so, but they are only able to do that by dropping way down the brook, close to the confluence with the Test. However, if that territory was 'taken', any incursion by Kuschta would be fleeting, so since they seem to travel down there on a regular basis it makes me think they have claimed the whole brook catchment as their own. Out on the main river, which is a bit off my regular patch, it is harder for me to know for sure, but from everything local river keepers report there are three or four additional family groups at most; more than that and Mion would be spreading himself too thin. It would be tempting to describe these bitches, some with pups and some single, as Mion's harem, but that doesn't quite take us to the nub of the relationships – it's more complicated.

Since there is no breeding season as such, it seems reasonable to assume that Mion's fathering duties were sporadic. Assuming that each female is required to be mated with once at a random time each year (probably less often, but let's keep the maths easy) and that Mion has four women in his life, then since the courtship and mating takes a week at the outside, that leaves him with the remaining forty-eight weeks to fill. So what does he do? Well, as we know, he is far from being a hands-on father. In fact he is pretty well totally absent. There are tales of otters who make brief, paternal visits to the litter and likewise others who watch from afar, but ultimately the mother will always see the male as a threat rather than a friend. There are far more tales of mothers spitting and barking at males who come close, and they will not hesitate to launch vicious attacks. Interestingly, the males don't fight back; they could easily overwhelm the smaller female but it would hardly be in their interests to make orphans of their offspring.

Therein lies the contradiction of Mion's life; he is seen as the aggressor by those he seeks to protect but he still has a duty to protect them. Now all that said, his duties over the summer were negligible; for male otters this really is a time to kick back and take life easy. The fact is that during the months of plenty there is not the same competition for food and, by extension, space. Mion is able to be magnanimous with interlopers when the mood takes him, as one languid twenty-four-hour period merges into another. The daytime will be whiled away on a comfortable couch; not Mion's per se, just one he happens to like – maybe deep in the reed beds or on a soft, rotten willow log. He is happy to 'borrow' whatever comes his way. Actually he is something of a creature of habit, plotting his journeys from one regular haunt to another. Lying there in the sun (it is always in the sun, given the chance; no damp, dark holt for him), he soaks up the rays from dawn to dusk as the sun tracks across the sky.

Otters might seem unlikely fans of sunbathing, but when you are designed with a terrifically high metabolic rate (it is 50 per cent greater than an equivalent land-based mammal) you burn off energy quickly. That figure gets even worse when combined with time in the water, which sucks out your body heat 27 times faster than air – remember, otters have no body fat to speak of. So, if you are able to stretch out in the sun, life gets a whole lot easier as you burn less energy. And in burning less energy you need less food, so there's no need to immerse yourself in chilly 51-degree chalkstream water. It is a virtuous circle. No wonder life is so good.

And what do they do on that couch? Well, in truth not very much. Like a cat, Mion sleep-dozes his way through most of the day, but being an otter he even manages to be busy when at rest, changing position many times with each hour. Mion is at his most sedentary when he curls up, his nose buried under the base of his rudder which wraps around the body, eyes tight shut. From this his particular favourite move is to uncurl, stretch, stand up, pad around the couch, executing two or three tight circles, before pawing at the bedding to settle down again, but this time curled in the opposite direction. I sometimes wonder how awake he really is when doing this, as the eyes barely open at all. This seems to be his most somnambulant time, with him content to stay this way even if it starts to rain, the water trickling down his fur, dripping off those long, pointed, outer hairs like so many gutter heads. Other times Mion crouches, back legs gathered beneath his rump, front paws before him and, if it weren't for the head laid half sideways over one of the paws you'd almost describe him as Sphinx-like. This is definitely an in-between posture; the stance that tells me he is more awake than asleep, ready to bound off when something catches his attention or the mood takes him.

But it is when he stretches out on his back, his rear legs extended, flopping over the edge of the couch, his front paws held in a begging posture just beneath his chin, that he looks truly at his most relaxed. His whole body goes limp, as if he has not a care in the world, the soft pale underbelly fur soaking up the sun. I'd assume in otter terms that this is the most vulnerable he'll ever allow himself to be, because at the slightest noise, or even the flitting shadow of a bird passing in front of the sun, he'll flip up in a trice, erect and ready for whatever, real or not, might be coming his way. Of course, more often than not it is nothing, but I guess Mion can't afford to take chances. But it's interesting that he tries to cover up these false alarms with a bit of grooming. Goodness knows who he thinks might be looking, but somehow he seems to be sending out a message that says 'Hey, I wasn't really startled, and anyway, that bit of fur really does need some attention'. He's fooling no one.

When he's not awake or on the move, in common with every other otter, Mion is a compulsive groomer. It is an oddly disorganised process; one moment he'll be chewing vigorously between his toes, the next twisting around to lick his back fur whilst in between stopping for a few seconds to scan his surroundings, then he'll scratch beneath his chin with the claws of one or other of his rear paws. There shouldn't really be any great surprise that Mion lavishes such attention on his fur, as it is his passport to a healthy life, but the summer can prove problematic with two particular parasites – ticks and fleas. You might wonder where on earth an otter might pick up fleas; they are not the sort of problem you might associate with the healthy, outdoor life by a river. The culprit is, perhaps unsurprisingly, those rabbits, which have a particular flea, the rabbit flea *Spilopsyllus cuniculi*. Not only are these fairly unpleasant for the rabbit, living in the fur around the face and ears, but they are also a carrier of the myxomatosis virus. In the days when myxi rabbits were commonly seen blundering about, blinded with puss-filled eyes, their heads would be crawling with these dark brown parasites, which are, in a bizarre twist of evolution, totally reliant on rabbits for survival. It is one of the strangest things, but the *Spilopsyllus* female must feed off the blood of a pregnant rabbit before being able to lay her eggs.

I guess that must be something of a relief to the otters, because at least they don't become hosts to the many fleas they'll pick up during the course of a rabbit-eating summer, though grooming out these tiny critters is a task in itself. They are no more than a millimetre in length, at rest the flea barely looking like a living thing. You'd almost be tempted to think they are a speck of dirt. But, replete with blood, they go shiny, almost black, and once that sliver starts to wriggle they are unmistakable. Actually, it is not entirely fair to blame the rabbits for the fleas. It is commonly thought that fleas require the warmth and blood of a host to survive, but not *Spilopsyllus*. For months on end they will happily survive in dry grass, just waiting for a ready host to brush past, a strategy which, believe it or not, has a particular name: questing. And in this questing they share a common cause with that other great pest for otters, the tick.

There is one tick that has a predilection for otters: *Ixodes hexagonus*, which is distributed by the hedgehog, hence its more common name, the hedgehog tick. In its normal state – before it has filled itself up with blood – the tick looks very much like a spider, and that's the family they are related to. Ticks start their one-year life as eggs laid in the soil, hatching into nymphs that go through two moults, shedding one outer skin for new growth before becoming that arachnid creature with eight legs. But it can't jump or fly, so it crawls up the stem of a plant and waits. Bearing in mind the vast extent of Mion's perambulations, it may well be in for a long wait, but there it is, poised at otter shoulder height with hooked legs, ready for Mion to come by. And when he brushes past, the patient tick grasps onto his fur at precisely the right moment. Contrary to popular belief, ticks don't instantly latch onto the nearest blood vessel; they take their time crawling around to find the best skin for their purposes, whilst looking for a mate on the way and using the otter fur as something of a love nest. That done, the female tick (male ticks are not heavy feeders) makes her way to the head, particularly to the ears where the fur is least dense, and where she will latch on to a blood vessel.

There is no great pain associated with the ticks for otters. They are more of a nuisance, but Mion will dislike them enough to try to rid himself of them in the early stages – you'll often see him lying on his back on hard ground or up against the bark of a log, wriggling and writhing as if he has an itch to scratch. Even if he doesn't shed them early on, the ticks don't hang around for long. Undisturbed, the longest they feed for is 5 – 7 days before they drop off to the ground, but it's unlikely that they'd remain on an active otter for that long. However, they can be numerous. A dozen at any one time is common, though some otters have been recorded with a hundred or more. The good news for otters is that ticks are seasonal, so by the autumn they will all have gone – unlike the fleas, which are a year-round pest.

The minor inconvenience of ticks and fleas aside, it's generally an easy summer for Mion. Food is there for the taking whenever he needs to eat, and when it comes to territory he doesn't really have to do anything to assert his dominance beyond simply being around. Marking out the boundaries of his kingdom with spraints is enough to ward off intruders from the outside; inside the boundaries the same scents will reassure those who rely on his protection that he is still around. As long as he covers all his ground every week or so, that should be enough, which is why, I suspect, I see so little of Mion around this time. The mill is one of the most far-flung outposts of his empire, almost the uppermost point of the river valley he travels. Any unwelcome otter would have been chased away long before he reached the mill. I guess, even though I missed his company, the fish in the lake had reason to be grateful for this summertime interlude.

That said, Mion's territory, winter or summer, will always be challenged in one way or another. The linear habit of travel that otters follow gives them little choice but to step in, out or through places to which they might not belong. This in turn asks questions of Mion you might not expect. Repelling insurgent males might seem an obvious point of conflict, but as you will have gathered, all the evidence is that otters really don't choose to fight unless they have to do so. An insurgent male in a determined search of a mate, well, I don't think there is much doubt that a fight would ensue, possibly with fatal consequences. If Mion is past his prime (it is going to happen one day) or has picked the wrong combatant, a new order will emerge. But one simply passing through? If times are good, a snarled warning will be enough to make the point.

Females in search of somewhere to live are a whole different thing. Here Mion has a nuanced judgement. He has to weigh up how much pressure his kingdom can take. Is there enough food and space for an extra family? Is one of the current mothers past her prime? Is a new bloodline required? If any of the above criteria fit, she'll be welcomed in, otherwise she'll be sent on her way.

Strangely, it is Mion's own offspring that will soon prove to be the most troublesome – maybe that's no surprise, as he has somewhere upwards of twelve of all ages and both sexes ranging around his territory. The bare fact is that, in the long run, none of these can stay. And for Lutran that clock is ticking faster than for the rest. So as the autumn equinox passes, the nights now longer than the days, the countdown to his eventual expulsion from the Wallop Valley has begun.

CHAPTER 10

IT'S A DANGEROUS LIFE

Frogs for lunch

Life is a dangerous affair for otters. In many ways you wouldn't really expect it to be that way; after all, they are big, strong and tough, masters of their particular universe, but it's true that the wild otter has a very short life. In fact, it's surprisingly short – if you make it to your second birthday you have beaten the odds, and only one otter in three makes it past that landmark and on to sexual maturity.

From the very outset the deck is stacked against you; infant mortality in the natal holt is high. Nobody has been able to put an exact figure on how many die in those first two months, but mothers like Kuschta are prepared to be brutal with runts in a litter. Postmortems regularly find dead babies in the stomachs of lactating females killed on the road, but they shouldn't shoulder all the blame; dog otters have been known to destroy entire litters. It is easy to put this down to animal blood lust, but maybe it is more complicated; a genuine attempt at territorial population control or natural selection. Of course there are all the other dangers from the world at large − foxes, red kites, owls and mink, to name a few − but for the young otter the greatest danger lies closest to home. Infanticide is a way of life in the otter kingdom.

You might think that, having made it past your first birthday, your chances of survival would increase. Think again. The odds get worse, this being completely the normal run of things in the remainder of the mammal kingdom. Usually death rates are high in the young then low in the adult. Not so with otters. The older you get the higher your probability of death. Which begs the question of an animal with no natural predators: what does kill you, then?

In one word: traffic. There is nothing in the DNA of an otter that gives it the ability to work out the peril of cars. Maybe a few more centuries of evolution will provide the tools, but for now in populated areas like southern England the roadkill death rate is as high as two-thirds. It is not really anyone's fault; otters move at night and with dark fur are hard to spot. Otters, like deer, are great creatures of habit, following the same paths for months, years, even decades on end, so you will sometimes see 'otter-crossing' road signs. Do they help? Well, maybe. Often otters have no choice but to go round obstructions: man-made weirs or a bridge that is blocked by a river in spate are typical, so a quick dash across the road is the only option for creatures that are always on the move.

All that said, the statistics might well be skewed, as they

are largely compiled from a corpse-gathering programme where members of the public reported dead otter finds, and they are, not surprisingly, mostly called in from well-travelled highways. So, cars aside, it is likely that otters can fall foul of many other forms of deaths that are hard to tabulate accurately. Today there are some things we may cross off the list for sure, such as hunting with hounds, which was made illegal in 1978 and stopped immediately, unlike, say, fox hunting and hare-coursing which were banned in 2004 but continue in various guises. There is no doubt that otters are shot and poisoned in areas where they are considered a nuisance, but by the very nature of the activity it is impossible even to guess at the numbers.

They also die by the hand of man in other ways, albeit accidentally – fyke nets, crayfish traps and lobster pots taking an annual toll. Fyke nets are rather like long socks with a wide-open mouth at one end, tapering to a closed-off point at the other. The idea is that, when fixed to the river or sea bed, the fish swim with the current, accidentally entering the net via the open end, but, once carried to the narrow far end, they can't escape, eventually dying of panic, exhaustion or most probably a combination of both. The same thing happens to otters. Occasionally they will escape – they have the advantage over fish of claws and incisor teeth – but modern nylon nets are tough and with just a few minutes of breath it doesn't happen that way often. Again it is an unanswerable question as to how many die this way as fishermen are disinclined to report incidents for fear of public opprobrium. Lobster pots are another potential trap – for sea-going otters there is very little more tempting, drawn not only by the smell of the bait (generally a bag of dead fish) and the vibrations set off by the lobsters as they try either to get out or to do battle with each other. It is relatively easy for an otter to push its way in via the spout, the tunnel-like entrance, but, once in, there is no escape. Fortunately most lobster potting takes place at a depth beyond which the average otter will not venture, though in the Scottish Shetland Isles drowned otters have be found in traps set in water as deep as 50 feet. It is fair to say that if you want to see coastal otters, follow the lobster boats and, at night, when the pots are piled up on a deserted quay, there are few better places for spotting a scavenging otter.

Crayfish traps laid in rivers are potentially more dangerous than lobster pots, even though they are smaller and lighter, because they are set at a depth that any otter can easily reach. Usually made of mesh, plastic or wire, they are one-yard-long, round tubes about the diameter of a football with entrance spouts at either end. Tethered by a cord, they are baited with anything dead and/or smelly (plenty swear by a punctured tin of cat food) and then sunk to the river bed. Scavenging crayfish will soon find their way in, and within twenty-four hours, if you have set your trap well, it will be full. Of course otters, again drawn by the smell and the activity, are compelled to investigate. You'd have to be a pretty determined adult otter to get into a crayfish trap (they can be fitted with otter guards), but sadly it is all too easy for younger, smaller otters who, like their sea-going counterparts, once trapped will drown within a few minutes. However, it is not all one-way traffic; once in a while I receive news of a trap found trashed on the bank. Human poachers usually get the blame, so I'll nod in dutiful acquiescence to deflect the blame, but most times I'll have a good idea from the detritus and damage that in fact an otter has, by accident or design, dragged it from the river and taken full advantage of an unexpected booty.

All in all, the death rate for adult otters is really quite alarming; one-third of the total population will die each year. In raw statistics, supposing the Wallop Valley and the section where it joins the River Test has twelve two-year-old otters today (which it probably does), then looking forward that means that only five or six will be left in a further two years' time, with just one making it to an eighth birthday, at which point the death rate ratchets higher still. Ten seems to be the oldest age that otters reach in the wild, though there are examples of 15–20 years when living in captivity. If you are wondering how they are so accurately aged, it is through the teeth, which have rings much like a tree.

So, aside from accidents, what is killing otters with such regularity? Well, disease doesn't seem to be a big factor, at least not in the wild. You might have thought that, being closely related to badgers and sharing a common habitat, tuberculosis might play a part, but to date there have been no recorded cases. In fact, there are no diseases that otters seem to be prone to catch that cause significant mortality. Occasionally there is an epidemic in other animals, such as the distemper outbreaks in seals during the 1980s, but, like TB, it didn't transfer across to the coastal otter population. In captivity it is a different story, where they are very prone to gum disease, gastroenteritic bugs, fungal infections in feet and tails, plus parasites in the fur. Most are easily cured with antibiotics, but it shows that otters are not best suited to living life in close confinement.

Ultimately otters, when not killed by cars, mostly die from themselves and nature. Though they try to avoid conflict, it happens, and though the fights are rarely to the death, they are a serious business. After all, you don't just have to cow your opponent; you generally have to drive him (or her) away entirely. This is not just about establishing a pecking order but also about preserving the food stock in your territory. So that pair of sharp, curved canine teeth on the upper and lower jaws will puncture skin and muscle with a purpose. It must also be added that domestic dogs play a part, with one-third of all bites identified in postmortems attributable to them. But it is not the wound that kills. It is the infection that follows, with somewhere between 10 and 20 per cent eventually succumbing this way. And succumb they often will, for as an otter you really do have to be in your prime to survive.

Life in and out of the water is draining. The otters need regular food but nearly every scrap is hard won. They have their status to protect. Long distances to cover. And if they falter they'll soon be trapped in a spiral of decline. Starvation is the biggest natural killer and it takes many forms: an injury that prevents hunting, old age, the inexperience of youth, competition from other otters, or simply the lack of food. It is a small consolation, but the transition from healthy otter to emaciated corpse is probably short. With few fat reserves to draw on, an otter soon loses condition. The protective natural oils in the fur dry up. The parasites take a hold. Wounds that would otherwise heal, fester. Speed and agility wane. The body starts to turn in on itself, with internal haemorrhaging, until eventually, sapped of the will to live, the otter has no option but to simply lie down and die. So one way or another, in coming up to five years of age Mion was on track to being something of an elder statesman. Kuschta, at three, plus one litter, had also outlived most of her generation. A second litter would certainly make her unusual. As for the pups, well, one was already gone and the odds were heavily stacked against all of the remaining three becoming parents themselves. The winter ahead would be a test.

The thing I notice most about transition from autumn to winter is not so much the denuding of the countryside but the lack of noise. That background hubbub, the soundtrack of the spring and summer, gently fades away until, quite suddenly, it is gone. The songbirds have mostly departed for sunnier climes; no longer is the arrival of dawn prefaced by a cacophony of competing voices. A few blackbirds try to fill the void, but their whistling caw song sounds rather plaintive, echoing around the bare hedgerows. The bumble bees, so noisy for so long, cease their daily quest for pollen as the worker bees die in their droves. Some queen bees are more fortunate; they head for ground, digging into a dry, well-drained north-facing bank where they will hibernate until as late as May. Hiding away is their protection mechanism, keeping them safe from the extremes of weather, predators (they like to avoid hedgehogs, badgers, birds and spiders – amazing how many things will eat a bee colony) and food shortages. Relatively late in the spring the queen will emerge (a north-facing bank warms last, so no false starts), seeking out the nectar in flowers on which she gorges to replenish her starved body. That done, she needs to find a place to nest. Abandoned water vole burrows are popular, so don't ever be surprised to see bees rising up from a hole in the ground if you are walking along a river bank. In this she lays her eggs, which in turn become larvae, which by further evolution spin a cocoon in which they grow to become adult worker bees.

At this point that strange bumble bee society emerges. At the centre is our queen bee. She will not leave the nest, but remains underground, cossetted and fed, constantly laying new batches of eggs whilst directing her workers, nearly all females, to guard or clean the nest whilst the others head out, returning with pollen and nectar to feed her and the next emerging generation. On this goes through the summer, until the queen senses that the colony must reproduce before the autumn sets in. So all of a sudden she starts producing males and queens who, once adult, leave. It is those new queens which, after mating, will hibernate over the winter whilst the original queen dies without ever seeing the light of day again.

I don't think I'm alone in noticing the silence. It even seems to seep into fur and feather. The fox, who a month earlier bounced along the tracks, tail sprung, head up and ears pricked, now slinks around with furtive glances left and right as he heads quickly from one bit of cover to the next. Maybe the start of the hunting season has him cowed; the rabbits don't linger long above ground, staying for as much time as it takes to grab some grass, fleeing for the warren at the slightest sign of disturbance. Even the heron, a slow mover at the best of times, manages to stay immobile for longer, as if stretching those yellow scaly legs a single step further is just too much effort. The hawks, usually so erect and proud perched on their vantage points, look forlorn, hunching up against the weather. In the summer they are constantly on the prowl but now I'll see them in a favoured tree top or atop a gate post for hours on end, the rewards of hunting clearly insufficient for the required effort. Some mornings they are grounded like aeroplanes at a foggy airport, as fidgety and annoyed as the passengers in the terminal, as the dawn mist shrouds the fields. It is a bonus for the field creatures who scurry about beneath this cloak of safety, but once the wind gets up the fog blows away and the meadows open to the hunt again.

There is one particular buzzard who has grown up around the mill; I've watched him go from scruffy fledgling to full-grown adult, who has now claimed this particular hunting ground as his own. I'm not really part of his plan, so on my walks he'll track every pace of my approach, as if daring me to come closer. But hawk eyes reveal nothing about the bird. Scruffy (I still think of him that way) meets my gaze but shows no emotion, those jet black pupils staring out from the yellow irises, unblinking with a slow turning of the head to follow my approach. He has to do it this way because birds of prey can't roll or swivel their eyes. Occasionally I will take the hint, turning back or changing my direction, but usually I don't. Why not? Well, part of me likes to defy the defiant, but mostly it is the chance to get so close to something so wild, even though Scruffy bristles with quiet menace. The beak, the talons, the eyes, the powerful wings – he is a killing machine that lives life by preying on the weak and unsuspecting. I'm not sure why that draws me closer but it does, until eventually he'll give in with ill grace. With a last defiant stare I'll see him brace himself by slightly crouching down, before thrusting skyward from the legs as the wings flap to give just enough lift for flight. And just enough it is. There is no sense of urgency or fear. Make no mistake, Scruffy is saying, 'I'm leaving under protest', expending the sheer minimum of effort to alight on a new perch not very far away. At which point I feel guilty for putting him to the trouble, but I'm sure someday soon we'll go through the whole performance again.

If some, like the rabbits and foxes, are starting to hunker down for winter, our otters are most definitely not. Though it might seem counter-intuitive, these last few months of the year are a time of relative plenty. If you are ever to find an otter starved to death it is far more likely to be in March than December. The fact is that many of those creatures that otters rely on most for food – fish, eels, frogs, rabbits, crayfish – are all at their most populous now because the offspring of the spring have become the adults of the autumn. Of course, a few months of winter will start to take its toll, but for now it is open season, though Mion, Kuschta and the pups will be wise enough to exploit some food sources before others.

If I had to bet right now which sole member of the trio will be alive a year from today (I'm not being dramatic; remember, the odds favour that outcome), my money would be on Wisp. It is not so much about the way she looks but the way she acts. Around this time (they will be eight months at Christmas) all three are well on the way to looking like adults. Wisp and Willow are two-thirds of their eventual fifteen-pound weight. Lutran, though noticeably bigger in frame than Kuschta, is still not quite as heavy as her. They are in all but one important respect perfectly formed otters, running, playing, swimming, relaxing and sleeping in one melded group. If it weren't for the difference in sizes you'd suppose they might, plus their mother, be quintuplets. But when it comes to the all-important hunting, things look very different.

Eels are a good point in question. The favoured food is on the wane, fewer migrating to the ocean with each passing day. The peak of high summer has passed, so they become an occasional treat rather than a regular meal. Kuschta, who seems increasingly frustrated with the reliance the pups have on her, still chases down every single one she can find as they are too nutritious to pass up. If she doesn't keep it for herself, she simply abandons it for the others; no longer does she bother biting the eel into sections. For themselves the pups still haven't the skills to catch them with any regularity, but, of the three, Wisp has the edge. In the river, eels are best caught by stealth, grabbed as they swim past a watchful but motionless otter. Wisp is good at this, waiting patiently in the shallows, beside the narrow channel through which she knows the migrating eel must pass. Lutran has no time for such subtleties, joining his sister for only a brief part of the vigil before wandering off, only returning in an attempt to steal, or at least encourage her to share, the catch. That never happens. They are past that stage. The best Lutran can hope for is a venomous bark of warning, spat from his sister. And Willow? Well, she is still largely tied to the apron strings of Kuschta, watching her every move but doing very little to catch her own food, still relying on her mother. But all is not lost for Willow; at this stage in her life it is normal for her mother still to be catching four-fifths of her food. It is an extraordinary level of dependency.

There is time enough to hone those skills; some of the things that are hardest to catch and that will survive all through the winter, like the trout, are, in a relative sense at least, left alone in these early months of winter. For now, the otters can concentrate on mopping up the creatures that will soon disappear from view, and so it is not a good time to be a frog. Otters have form when it comes to frogs; they are one of the few things that they actually play with, like a cat with a mouse. From an early age the pups were fascinated by the frogs that at first had no problem hopping out of harm's way. But as Lutran, Willow and Wisp grew older, the balance of power changed as frogs became a daily dish. Frogs might seem an odd food choice for otters; at first glance, there doesn't seem to be anything particularly appealing about them. A bit slimy, often small and maybe tricky to catch, but I guess some of us eat frogs' legs? But they are, along with toads to a lesser extent, really important to otters, consisting of at least one-tenth of their general diet, rising to half in critical times. Bearing in mind how close their respective lives are entwined, maybe it isn't so odd after all.

There is really only one frog species in Britain, the not very romantically named common frog. Its Latin classification, *Rana temporaria*, gives the species a little more gravitas. Ranid is of the frog family, so if you suffer from a fear of frogs you are a ranidaphobiac. I'm guessing that the *temporaria* suffix refers to the life cycle of our *rana*, but I am not entirely sure why, as they live a surprisingly long time. There is one other British frog, the pool frog *Pelophylax lessonae*, but this is largely extinct in the UK bar a few colonies in Norfolk and some other isolated spots. For the most part frogs like to inhabit exactly the same space as otters – damp meadows, ponds, ditches and the margins alongside the river – so it's no wonder they have so much in common with *Lutra*. Frog breeding time is a positive bonanza for the otters, as the courting couples like to congregate in one place, where the males compete for females by producing the longest and loudest calls. The winning male gets to mount his female, fending off suitors with his hind legs, as he fertilises the two thousand odd eggs that we more commonly know as frog spawn. These will then float in clumps (toad spawn is laid in strings) until they become tadpoles, before eventually meta-morphosing into tiny froglets after about three months.

As you might imagine, lovesick frogs, distracted by other things, make for easy food, so in the early spring, in the weeks running up to the birth of the pups, Kuschta took to hanging out beside the ditches in The Badlands, picking off frogs at will. No hint of ranidaphobia in her for sure. The hunting was not hard, for it is the misfortune of frogs that mating takes place at night, so she'd simply swim up behind a distracted couple and with a single bite often get two for the price of one. I'm fairly sure that the first bite does for the frogs and it takes a matter of a minute or two for Kuschta to chew them then gulp them down. Then she'd slide back into the water, circling round for the next snack. The remarkable thing is how little spooked the frogs were by the appearance of Kuschta and the subsequent disappearance of their colleagues. Kuschta would surface, plucking one or a pair, whilst another pair, almost immediately adjacent, would continue about their business totally oblivious to, or perhaps simply ignoring, the fate of their neighbours.

You'd think that the frog spawn might also be a tempting target, as the combined frog population produces it by the barrel load, but apparently not. Despite the fact that the eggs are full of protein, there must be something in or possibly on the spawn, as not only do the otters never touch it when floating in water but they carefully eat around the eggs inside their captures. So occasionally you'll find little piles of frog corpse debris mixed in with spawn. In fact, there is something of a debate about how otters go about eating frogs. It is a commonly held belief that they skin them, but I have never seen this happen and, frankly, it would be a mighty fiddly affair. It seems more likely that rats (as opposed to water voles) are the culprits of this particular habit. Crunch, chew, swallow and then move on to the next one is the otter method; after all, they'd need to eat twenty or thirty to fill up for the day.

But what is true is that otters, when they have a mind to, persecute frogs unlike any other creature that comes into their orbit, and by the early winter, as the population peaks prior to hibernation, there were plenty around to amuse Willow and Wisp. Frogs actually spend less time in water than might be supposed. As juvenile frogs, newly hatched from their tadpole stage, life is very much spent in the water, but as they become older they spend more and more time on land. By the age of two or three years (frogs can easily live to eight years) they rarely leave dry land except for mating or hibernation. Their favoured foods – ants, insects, slugs, snails and worms – are far more readily found amongst the tussocks and grasses, which, despite their brown and rather bedraggled state, are still seething with wildlife.

As a general rule frogs don't hop out in front of Willow and Wisp, offering themselves up as an easy meal. The pair have to search them out. The frogs like best to live in that space between the earth and the fallen grasses. It is damp without being wet, largely sheltered from the worst of the weather, home to all the things they like to eat, and safe. Well, not always. The pups soon perfected a reliable hunting technique that involved inserting their heads between the earth and grass, pushing forward rather like a snow plough and parting the grasses as they went. Sometimes they became so buried that the best you could do to track their progress was listen for noise and watch for the movement of the under-growth. How they track down the frogs in the first place I'm not entirely sure. I suspect smell is what draws them in the right direction initially, but once the frog home is disrupted by the arrival of Willow or Wisp (Lutran was always far keener on bigger prey), those instincts for vibration and movement, that work so well underwater, play out in exactly the same way on land as the frog tries to flee.

It is rarely an instant result; the pair sometimes collec-tively, but more often individually, will weave through the grasses head down, boring forward, those sensitive whiskers picking up the vibration of the fleeing frog until it is cornered or makes good an escape. If it was the latter, an otter head would pop out from the grass, sniff the air and maybe pause for a brief groom (which feigns indifference to failure) before resuming a new hunt. If successful, one of two things would happen – eating or playing. Eating is swift, playing is not. Why on earth otters pick on frogs I have no idea, but it seems to give them particular pleasure. Some say that frogs have a fluid in their skin that otters find unpleasant, but this is excreted when the frog is stressed, so if the playing goes on for long enough the taste will be gone. Whatever the reason, once cornered, the poor little thing will lie shivering, pressing itself into the ground, as the otter nose gets closer and closer, until it is forced to leap for safety. Of course, there is no safety, as neither Willow nor Wisp has any problem bounding along with as many leaps as the frog cares to take. Occasionally they will use a paw to bat at it mid-air or even leap to grab it in mid-flight, but eventually they do the decent thing, extinguishing the life of the terrified creature.

The odd times the frog makes it to water, a whole new game begins, the otter playing contrarian: frog swims towards the bank, so otter pushes it further out. Frog swims out, otter pushes it to the bank. Frog crawls up the bank, otter slides it back into the water. You are getting the idea – the frog can't do right for wrong, and sometimes an otter paw, both delicate and firm, will simply pin it to the bottom until it is allowed to re-surface. Willow and Wisp will make a real meal of the whole affair, and what is bizarre is how nonchalant they like to pretend to be, turning their back on the frog as if to offer some chance of escape, even to the extent of giving it a yard or so start before reeling it back in. But it is in the end all pretend, and when they tire of the escape antics the frog simply becomes food.

All that said, frog hunting is not always a frivolous affair; in the depths of winter when the amphibians hibernate in the muddy margins it will be far more serious. January and

February is the time when they disappear from our view, sinking themselves down in mud where they go pretty well comatose, their metabolism near to death-like. To all intents and purposes they are inert, breathing by absorbing oxygen through the skin. The usual clues for finding them – smell, noise or movement – are denied to the otters, but they will not be deterred. When food is at a premium too much is at stake, so they add a new weapon to their hunting armoury – blind touch. It is simplicity in itself, albeit a little messy, and Wisp was to become a past master at it.

I suppose frogs never really expect to be disturbed when hibernating. After all, what could be safer than being submerged in mud? Well, apparently plenty of things because being covered by a few inches of swamp is no protection from a determined otter that is trawling the mud, wading through the wet margins of the meadow ditches and the brook. Wisp creeps forward slowly, sunk to her waist, pausing with each shuffled step to balance on three legs whilst extending the fourth leg out ahead of her, the digits of her paw stretched wide. Pushing down into the mud, she moves her frog-finder paw in an arc like a mine hunter, feeling for soft flesh somewhere in the boggy morass of mud and water. Of all the hunting they ever do, this is probably the most measured; methodical and thought out. It is certainly effective, judging by how regularly Wisp finds her prey. You can see this the moment she touches something alive; the leg stops, stiffens, she pressures down and then all it takes is a flex of those claws to snag the sleepy amphibian, who probably never awakened to his fate.

In this particular world many must die for a few to survive.

CHAPTER 11

SOLSTICE

The end of the beginning

A few days before Christmas the midwinter sun sets between the two upright stones of the great trilithon at Stonehenge. Up on Salisbury Plain the modern pagans are gathered to celebrate the solstice, that moment in the celestial calendar when the night is at its longest and the day at its shortest. As the last rays of the sun bounce off the lichen-covered sandstone, the famous monument silhouetted against the empty landscape, it is tempting to wonder why a civilisation four or five millennia ago would have created this megalith to mark the depth of winter. How could that be a good thing?

Well, pagan or not, I guess they would have appreciated the words of Churchill in November 1942 when the tide of war was turning: '*Now this is not the end. It is not even the beginning of the end. But it is, perhaps, the end of the beginning.*' For that moment of the hibernal solstice told ancient civilisations that there was hope. Winter was not eternal. Spring would arrive again. Life would be better once more. It is easy for us, with all our modern knowledge, to dismiss such events as trivial, but for primitive man the dividing line between life and death was marked by such transitions.

In the natural kingdom this solstice may not have a calendar date, and for sure Mion, Kuschta, Willow, Wisp and Lutran don't gather in appreciation of its happening, but the implications are equally profound. The next four months would be the most testing of their lives, both for the pups and the parents. Do they know this as they enjoy the longest night of the year? After all, darkness is the otters' friend. I suspect not. Animals have instinct, not knowledge, so they were not to know that each passing month until spring would become daily more difficult. It would be not a process of peaks and troughs but rather a continual downward spiral. Survival will depend on many things as the forces of nature bear down on them all: weather, food, conflict, the protection from others – these and many other factors will determine who survives and who dies. To restyle Churchill, it is either the beginning of the end or the end of the beginning.

Up to this point they have all had it pretty good. The months that precede the solstice might be winter in name, but, as we have seen, it is not such a bad time to be a creature of the Wallop Valley. From my favourite vantage point high up on the ridgeway the landscape below maybe doesn't look so inviting to the human eye. Every tree, bar a few straggly pines, is shorn of its leaves. The hedgerows, trimmed and tidy, march out in all directions, now dormant until the spring. The ploughed fields are either bare dirt, speckled with shiny flints washed clean by the rain or covered with combed lines of winter wheat. The Badlands are universally drab; everything seems dead or on the way to being so. The once bright green reeds are fading fast, withered russet along the edges. The sedge grasses are dry and brittle. Even the bulrushes that have stood tall all season are starting to keel over. The velvety flower spikes, which look more like a hotdog sausage than a bloom, gradually disintegrate into a soft down that gets carried away on the wind until all that is left is a jagged stem. Even the Brook looks out of sorts, still rising from its summer low as it wends its way through a landscape that seems to have been stripped of colour by some unseen hand.

But come down from the heights, or just take time to pause amongst the gloom, and suddenly there is more to the landscape than meets the eye. The hedgerows may be dormant but they are larders for the winter. Clusters of red/black hawthorn berries are everywhere, as are the last of the blackberries. By now they are shrivelled, rock hard to the touch, but tiny ants and flies crawl all over them, extracting every last bit of juice, whilst spiders, spotting an opportunity, weave new webs each night between the bramble briars. Fat sloes, the same ones that go into gin, are all over the spiky *Prunus spinosa* bushes that make for good cattle-proof hedges. It looks so pretty in spring with its creamy-white flowers, typical of the plum family from which it originates, but as it fruits it becomes something of a menace, living up to its more customary name of blackthorn with dark bark and vicious thorns. No doubt it grows this way to protect itself from grazing animals; try to push your way through it and you'll soon see why, as the barbs will pierce any boot, let alone clothing. Menacing though the plant may be, the fruit looks tempting, just like oversized blueberries with that same dark purple hue, with a slight hint of a white bloom. However, bite into them and you'll discover the fruit is just as unpleasant as the thorns. The astringency of the sloe juice will shock you, the sour tannins drying your mouth to sandpaper in an instant. Even if you spit it out instantly you'll still spend the next hour making your saliva glands work overtime to bring back any sense of taste or feel, whilst your tongue will feel like a stranger in your mouth.

There are clearly plenty of better pickings. The ground around every hazel bush is surrounded by empty husks, each one sporting a tell-tale hole with tooth marks from a squirrel who has extracted the nut. And the field mice are working overtime scurrying around gorging themselves every day as the wind dislodges a new batch of seeds from the dead grasses and plants. Like the moles, rabbits, badgers, stoats and other fellow creatures, they are building up reserves, taking advantage of what is on offer, getting ready, hunkering down, laying in stores. But for our otters there is not very much that they can do.

The fact is that otters have no real ability to adapt to the changing of the seasons. Yes, they might undergo a winter moult, their fur turning a shade darker with a bit more protective oil, but that is about the total sum of their physiological changes. Their metabolism continues at its high rate. There is no hibernation for them, and they are forever caught in that requirement to consume 10–15 per cent of their body weight each and every day. Fat reserves, such as they are, can't be built up for times of austerity. Food can't be hoarded. Life can't be dialled down. Kuschta can't eat less so her pups can eat more. Mion won't come by to provide a helping hand. Lutran, Willow and Wisp won't suddenly become independent.

Knowing all this, you might suppose the winter weather to be the enemy of our otters, but everything I have ever seen of them tells me they really don't care. It can be belting with rain, howling a gale or the ground covered with snow, but they will appear at the mill doing exactly the same things in exactly the same ways. To some extent I think they might prefer dreadful weather; it gives them the perfect cover to go about their business, when their enemies and competitors, such as they are, are less likely to be out and about. There is, of course, the added bonus of all that extra darkness. In the height of summer they are restricted to as little as four or five hours of night; for creatures that like to hunt a bit, rest a bit and repeat that process a further two or three times before curling up to sleep for the day, that's something of a nuisance. But come the depths of winter when it is dark by four in the afternoon and not light until seven in the morning, well, that's fifteen hours to call the world your own.

And so they do. A typical night will start after an idle day in the holt, the outdoor couch life of summer abandoned in preference for the underground homes. I say underground but I've discovered, by accident rather than design, that they are not so much bothered by the subterranean imperative. The section of the Wallop Brook we look after for fly fishing takes a fair bit of management even though we like to keep it, as far as possible, wild and natural. But inevitably trees fall over in the wrong places, blocking the river, obstructing a path or just becoming plain dangerous. You get the idea. So the chainsaws come out and we chop up the offending tree. Now, in an ideal world I'm sure most people would take the logs for firewood, but we are long past dragging wood hundreds of yards across boggy ground, so we pile up the lumber in a triangular stack. As it rots down over the years it makes a great home for all kinds of grubs, insects and worms, and, as it happens, otters, who soon use the crevices as a handy home.

It is far from being a complete holt, but when the piles are close to the river they soon become regular haunts. The paw marks, spraints and rubbed smoothed bark are evidence of regular use, though I suspect the wood heaps are more of a halfway house, something between a couch and a holt. Anyway, I read somewhere that North American otters, similar to our own otters in many respects, often adopted abandoned beaver lodges, which, despite being partially in the river, are largely constructed above the water. That had me thinking: could we improve on the log piles to create a 'proper' holt without the effort of digging? The answer is, you can, so I thought I'd try.

The meadow immediately below the mill, divided as it is by the brook, is a regular thoroughfare for all the otters. On one side the land slopes down to the water, the grass grazed by sheep with a path for the fishermen. At first glance it doesn't look the most inviting route for otters, a bit exposed and open, but they do travel along the bank under the cover of night, especially if they want to move at speed, and by the end of the summer they will have worn quite a path of their own. The brown pounded grass is easy to spot, weaving along the same route the fishermen take, except for the occasional deviation into the river, where the muddy slide is a giveaway. Sometimes on one of the little promontories I'll find the remains of a rainbow trout. It is clearly a capture they have carried down from the lake so I guess they don't mind being out in the open too much, which may well be because the refuge to the undergrowth of the opposite bank is just a short dash away.

The other side is flat ground, not quite a marsh but the remnants of a traditional water meadow that feels soggy under foot, even at the height of summer. It is really not much used or trod, left to its own devices for most of the year to become a dense, grassy thicket. We don't do anything with these few acres until the end of the summer, and I think the otters know this, as I'll see plenty of paths from the river disappearing into the rampant pasture. However, it can't be left entirely alone as it requires some sort of management, so towards the end of the summer a small herd of cattle will be shipped in for a month. To start with, you'll barely be able to see them, so tall are all the wild grasses and plants; just the backs or the occasional bovine head will show above the confusion of seed heads and white flowers of the hemlock water-dropwort that thrives in profusion. Some sort of order will gradually assert itself as the cattle randomly graze, picking at the best of the forage until the rains come and it is time for them to go as they start to do more harm than good, churning up the ground. In one final play to bring order to nature's chaos we'll send in a tractor with a flail that will smash to pieces what the cattle deigned not to eat, the field turned a sort of silvery grey-brown, covered with the chopped stems, grasses and all the detritus that the flail scatters in its wake. For the next few weeks the meadow will become something of a killing field, as every vole and mouse exploits the huge bounty of seeds we have so helpfully gathered on the ground for them. You can even hear and see the piles move as the foragers scurry about beneath. Of course we have not entirely done them a favour, as their cover is largely blown by cutting. If I can see them moving about then for sure can the owls, Scruffy and all his fellow raptors that gather at their respective vantage points to swoop down when the moment is right.

Even after the flail the meadow is not entirely bare, for along the bank stand eight really old willow trees. How ancient I am not sure, but they have by all accounts been pollarded for generations and regularly appear in sepia photos from the early days of photography, with our flint tenth-century Norman village church and its squat-square bell tower providing the backdrop. Some of the trees have such girth that it would take three people, arms outstretched, to completely encircle the trunk. The word 'gnarled' really doesn't do them justice; the bark is rough, actually really quite sharp to the touch, irregular and jagged. Some sections are peeling away, exposing rotten wood beneath, the cavities so deep and so large you might easily slide your arm inside. You'd think this would kill the tree, or be the beginning of a long demise, but it does not seem so. Willows, or these ones at least, have remarkable powers of recovery; new bark will gradually grow over the wound until it meets in the middle. I have a feeling this might be Nature's form of a sticking plaster, albeit measured by decades, for if you tap the shell of the trunk in some places it definitely sounds hollow.

There was a time when these willows were important to the village, the pollarding done for a practical purpose. Today we largely pollard trees for aesthetics but in times past the willows provided a constant source of pliable timber. Essentially it worked like this: every few years all the branches were lopped off from the top of the trunk, the crown as they call it, in what is essentially extreme pruning. This promotes new growth and the willow shoots grow upwards, rather like an extreme spiky punk hairstyle, the bright new growth in contrast to the gnarly old wood of the ten-foot-high trunk from which it grows. The willow sprouts with extraordinary speed. After a year the withies, flexible whippy stems of ten or twelve feet long, are harvested for basket weaving. After another year or two, if left uncut, they will thicken to become useful for construction – thatchers in particular use them for the roof frameworks and pins to secure the thatch. A few years more and the long straight branches are perfect for fencing. There are, of course, myriad other uses, but I'm sure you get the general idea, and by rotating the pollarding amongst the group of trees the village was assured a constant supply of different lumber.

We still pollard the willows every decade or so but I can't pretend it is in any way, shape or form for the original purpose intended. Ours is really more practical. Once you've started pollarding a tree you have to keep doing it, as if left unchecked the tree becomes top heavy, and is eventually toppled, trunk and all, by high winds. So, chainsaws in hand, again we gave them a mighty haircut, and it was the debris from one of those occasional cuts that gave me the idea for a man-made holt.

In the floodplain of the Wallop Valley, close to the river, creating a holt underground, even in the height of summer, is impractical; dig down as little as a foot and the hole will be full of water in a matter of minutes. So, something above ground, assuming the otters will use it, makes perfect sense as well as being a good deal easier to construct. It is not as hard as you might think; first you need to find your materials – well, nothing is better than all those willow branches and brushwood, which, given a year since pollarding, are all nicely aged. No otter would suspect a thing. Now it is very possible to construct holts with multiple chambers and connecting tunnels to imitate the real thing, but I was starting with something simpler: a single sleeping chamber with an outer perimeter with two entrances, one facing the river and the other the meadow.

To start, I chopped into lengths some of the thick branches that were left from the pollard; these were to be the walls of the central chamber. I took the logs, placing them in a square on the ground to allow for a space inside of about two feet square. One log was a bit shorter than the other three to provide an entrance that faced upriver, as I was building this only a couple of feet from the bank. That done, I fashioned stakes from thinner branches, hammering them upright into the ground to hold in place a second layer of logs on top of the first, then repeated this, adding more layers until the walls were complete at about 18 inches high. Next I placed two longer logs of roughly six feet each in an L-shape along two sides of the square chamber, using more stakes and logs to bring what were effectively to be the entrance and exit tunnels to the same height as the chamber. Part of the trick here is to allow enough space for the otters to come and go with ease, but without making it seem too open. They need to feel safe inside, so a one-foot-wide tunnel is about right, letting them turn around if need be. One of the tunnels opens to the river's edge.

That was the basic structure complete; all it required next was a roof and some disguise, but first I gathered some dry reeds to line the central chamber, where they would sleep. I'm sure the otters will bring in their own bedding if they ever call my 'den' home, but a head start would not do any harm; it made me feel good anyway. The roof, a two-layered affair, is far more laborious than the walls, requiring dozens and dozens of straight poles that are laid side by side across the top not only of the central chamber but of the tunnels as well until, some hours later, the whole thing has a flat roof. This alone is not enough; the gaps between the poles let in far too much light. Otters like their holts dark. So with a spade I layered on turf, leaf mould and dirt, letting it fill in the crevices. I discovered this was not a one-time job; it was far better to return a few days later after the rain had padded down my work, giving me the opportunity to apply more dirt to the fissures that had appeared. I think it was by about the third or fourth visit that the roof could be declared fit for occupation. The final task was to take the remaining brushwood and pile it randomly over the top of the whole thing, and this helped it merge into the landscape. Truly, from fifty yards you'd think nothing of it. That, I thought, was good.

I'm happy to report that a year on from the great holt experiment, it is a success. I'm not entirely sure which of the family uses it; my suspicion is that at one time or another they all have. After all, it is said that in the typical otter territory there will be anything up to thirty resting places ranging from the occasional couch to the full-bore holt, so I'm sure mine passed muster somewhere on that scale. I can't say I ever saw them inside. A bit of me has wanted to peel back a section of the roof to take a peek in the sleeping chamber, as it would be interesting to see how they had bedded it out, maybe even catch a glimpse of a sleeping otter. But on the other hand, fiddling about might scare them away for ever. So I have contented myself with observations from afar, though in summer when the meadow is fully grown the actual structure is hard to see. From the far bank the slick, muddy grass patch that marks the short trail from river to entrance is readily visible. The regularly attended pile of spraints indicates ownership. The matted carpet of red willow roots that form a little platform on the water's edge is where they obviously sit to eat. Different days, different things – fish parts and red shards of crayfish are the most regular signs of visitors. Best of all, the holt is losing its man-made genesis – the willow stakes have taken root, starting to grow as new trees, and even the logs laid sideways are sprouting. It will never be entirely natural, but in time it will get close.

As they emerge from the holt at dusk, otters are not so different to us first thing in the morning, as sleep gives way to wakefulness. It is a time to stretch, groom and prepare for what is ahead, plus grab a bit of breakfast. After a night curled up, they like to roll onto their backs, extending legs and toes to their fullest length, momentarily going stiff as a board before relaxing enough to wriggle against the grass. You can see this gives them enormous pleasure; whether they are itching, easing the stiffness or simply like the scratching sensation, I do not know, but they seem to have a happy, stupid grin all the while, teeth just showing between those hairy lips. Then they pop upright and sit on their haunches, sniffing the air looking all about, eyes, ears and whiskers alert to everything around them. Then it is grooming – never an ordered affair. A leg comes up, claws out to scratch around the ears. Then the head goes down, rapid little movements of the teeth working at the base of the tail. Then some licking of the belly fur, before curling around to work the back. And so it goes on until, body buffed, our otter is ready for the night ahead.

Otters really are as happy on land as they are in the water. They may not be as graceful, but it is a more practical way to move around – every minute in the water burns off calories that are best conserved, and the truth is that rivers, even ones as productive as the Wallop Brook, are far from being continual larders with food evenly spread out along its length. For instance, the fast, shallow sections look inviting to the human eye and ear, gurgling and sparkling as the water rushes by, but for a fish of any size, well, it is no place to be. Holding station is simply too hard and too much effort, not to mention the difficulty grabbing any food that comes speeding by. Crayfish feel much the same; scavenging is better done elsewhere, and, in truth, along with the fish, they feel far too vulnerable to predators like herons. So it is really to the deeper, slower sections of a river that most river life heads, where survival is easier and, for the most part, their days are less hazardous. I say less hazardous because everything is relative; otters know how their prey think, for the most part skipping the unproductive shallows (there is an exception, but more of that in a while) in favour of the pools.

So along the bank our otter pads, seeking out the first pool of the night; if all goes according to plan, a fish will soon be caught and eaten in just as long as that takes, which is rarely very long before yet another bout of grooming. Now at this point the toughest choice our otter has to make is rest then move on later, or move on now then rest later. Staying put all night on that single pool is not an option, for reasons otters seem instinctively to know. On a purely practical level that first foray will have spooked all the fish, the remainder scattering, hiding themselves away. Seeking them out would be too much effort, the law of diminishing returns setting in. Far better to leave it be for another night or two. Let normality return. On that more instinctive plane, otters understand that they need to conserve food stocks; denuding a river of every fish is no way to guarantee their future or that of the fish. So, with a spraint to tell the tale of the evening to any later passing otter, ours moves on.

How far depends entirely on the river and the time of year. In the summer, when the water is lower, the fish tend to congregate in the pools, which will be fewer and farther apart. In the winter, with more water and more depth, they spread out, so as a consequence are harder to locate. The amount of water is not necessarily the determinant of how many fish any particular stream or river will hold. Some rivers, such as a chalkstream, rich with food, have a very high density of fish that might be measured in thousands per mile. Others, like a moorland stream, prone to extremes of flood and drought, very few. Otters don't travel great distances because they choose to; they do it only when they have to.

A couple of hours into the night, our otter will be resting up but starting to think about the next meal – that famously fast digestion system empties the stomach in not much more than an hour. Of course they have their favoured spots, couches and similar, but frankly in the dark and on a familiar river they are happy to rest up wherever takes their fancy. They don't sleep, but rather stay alert to every movement around them. After all, they have all day to slumber, and the night is about gathering food. But sitting on a bank staring into the dark is no way to find food. Otter eyesight is not brilliant at the best of times and it is often said that they see better underwater than on land, so in the quiet of the night it is hearing that provides the clues to what is happening out there in the darkness, where little is easily discerned.

I know from my own experience that otters have incredibly sensitive hearing. No longer do I carelessly let a gate latch click shut or step on a stick – they will be spooked in a trice. Weirdly, my torchlight will go ignored, as will I if I stand stock still. There doesn't seem to be that scent thing that will alert a deer at two hundred paces. Hearing, and the importance of it, is learnt early; in fact it is the sense that develops fastest in the natal holt, giving some indication of how important it will be in later life. So, if you are an otter out on the river at night, when your eyes and nose won't help you much, it is the ears that come to your aid.

If you ever have the chance to sit quietly beside a river

through the dead of night you'd be amazed how quickly your ears will become attuned to the subtlest of sounds, even though rivers are noisy things. It might be tempting to describe the bubble, sucking, tumble and flow of the ever-rolling stream as white noise, a constant background hubbub that drowns out all other sounds. Well, in a way that is true. I for one often become oblivious to everything except the river, sometimes my reverie broken in shock by a cow or sheep that suddenly appears at my side. But soon you'll get the cadence of the river. The unbroken hubble-bubble as water pours over a weir. A strange sucking slurp where the water folds over a waving raft of weed. The regular jingle as shallow, nervous water tinkles over stones. And then amidst all that you'll suddenly hear something different. Every other sound will fade away as your ears focus (if they can do such a thing) on both the noise and whence it came.

Sometimes you'll dismiss it quickly – a crack, then the splash of a branch that snapped from a tree. Other times it will be too distant or indistinct to identify. And yet in other times you simply just know it comes from something alive – a fish leaping from the water, maybe fleeing a predatory pike, the splash-landing echoing along the river, the glooping sound as a cruising carp arches its back and dives deep, the water closing over it. Or some frantic splashing as a spawning salmon forces itself upriver across the gravel shallows from one deep pool to the next. Whichever it is, you'll strain your eyes to discern what you've heard. Sometimes luck will be on your side, the moonlight catching the movement of body or tail, but more often than not all you will get is darkness until the next sound redirects your gaze.

Spend your time like this and you are doing no more nor less than our otter, though he or she is able to go one better.

For them a sound is the clarion call, an instant indication that there might be something out there to eat, or at least hunt. I'm sure they, like us, are able to distinguish between the various noises, choosing which is worth investigating and which is not. Actually they are probably many times better. After all, getting wet, losing warmth and all the subsequent grooming to get dry again is not something to be taken lightly. But if they do choose to investigate, they have a tool far more potent than any of ours. Slipping into the dark water, suddenly the vibrissae, their whiskers, relegate eyes, ears and nose to also-rans as the hunt begins.

It is easy to think of whiskers simply as long hairs, but they are more complex than that. Otters have evolved something that goes beyond what you'd typically see on a cat or dog, or even on their cousins, the badgers. What they have has more in common with seals, walruses and sea lions. Under a microscope the cross section of an otter whisker is an incredibly complicated thing, the central hair surrounded by flexible tissue, vibrissal nerves and several blood-filled sinuses that connect to the hair follicles, which in turn have receptors that transmit to the brain. All this gives otters an ability termed 'hydrodynamic trail following'. It sounds fiendishly complicated but it is simply that ability to detect and follow the trail of turbulent water formed in the wake of a swimming fish by using the whiskers alone. Simple it may be, but it is the most potent weapon for underwater, night-time hunting, and without it otters would probably not exist in Britain today. For as successful nocturnal predators they have managed to defy centuries of persecution by man, living not just under the radar but under the feet of those who would wish to do them harm, except for the fact that oftentimes diurnal man has been oblivious to their existence.

Hunt number two of the night should go the way of the first; a brief flurry of attack, successful capture and consumption. If not, our otter might hang around for a while if the initial commotion was not too great; after all, the fish are not gone, but have rather gone to ground, hiding out until they think the coast is clear. The fact is that in a lifetime of patrolling the same river and the same pools, otters get to know the habits of their prey; the bank undercuts, back eddies, tree roots and boulders in, under or around which they like to hide. Suddenly those same whiskers go from being movement sensors to touch sensors as the otter pokes its head into familiar hideyholes. If the prey bolts, it is game on. If it stays put, then the otter may deploy that rather clever bubble-smell technique to determine what exactly it has cornered, or simply press home its advantage without any further ado – I'm guessing that an otter doesn't have much problem working out when it has a crayfish or eel at the end of its nose. One way or another, our otter has plenty of ways of eking out a meal from most situations.

By now it is over halfway through the long winter night, well past midnight. It is fair to say that barely any other creature will be moving in the river valley. Otters reign supreme, hunting out the night until the first fingers of dawn signal the end of darkness, giving time enough to find the nearest holt to sleep the incipient day away. Some nights will have been good, others bad, but persistence has its rewards and at certain times of the year nature will offer up an unexpected harvest.

It might seem counter-intuitive, but winter is the breeding season for trout and salmon; that time when the return of *salar* from the ocean achieves its ultimate purpose and the native *trutta* casts off its usually indolent lifestyle for a month of frenetic (I use the word advisedly) activity. In choosing December and January, game fish, as trout and salmon are characterised, are different to most other fish you will find in British rivers. Pike, carp, tench, bream, perch – all those other freshwater species that are usually lumped together under the term 'coarse fish' – will do their spawning between March and June. The only real outlier is the grayling, a favourite of our otter, which, though generally thought of as a game fish, spawns in the spring along with the majority of British fish. Nonsensical it may seem to us, but for game fish winter, despite all its apparent deprivations, provides the perfect nursery for their offspring. If you are a trout or salmon ova, once you leave your mother you are on your own. There is no protection. No further upbringing. No concerned father in the guise of a bullhead. The best your parents will have done for you is dig a 'redd' – a shallow indentation in the gravel into which you will be laid. It is not really fair to dismiss entirely the effort a pair of spawning trout will put into creating a redd. It will take a few days, becoming something of an obsession as the two alternate between courtship and digging. They become oblivious to danger (you might sense where this is going), circling around each other, lying side by side or suddenly peeling off with a burst of energy to attack the gravel, flexing their bodies and tails to gradually gouge out a hole in a fast-flowing section of the river that is not much deeper than their own bodies.

Once the redd is ready the pair will relax for a while, comfortable beside each other at its head, just hanging in the water, gently holding station in the flow. Then quite suddenly the female will go rigid, the eggs pouring out from her body whilst the male shudders as if electrified as he sheds his milt. This mixes with the eggs on the current, and they tumble away over the redd. As you might imagine, this is now a fairly haphazard affair, with many more eggs carried away down-river than ever lodge in the hole in the gravel. But there will be enough swirling in the back eddy created by the redd that will hang there before slowly dropping down onto the bed of the cavity whilst the parents have time to make them safe, using tail and body to dislodge enough gravel to gradually cover them, backfilling the hole smooth and flat. Nestling safely in the gaps between the gravel, the eggs will stay there for the next two to three months. The fast-flowing water provides all the nutrients and oxygen they need to develop until they hatch into alevins, little fish with an egg sac, which continue to live their lives under the cover of the gravel until they emerge as parr, which are recognisable as tiny adults. It is this evolutionary need for trout and salmon to seek out those shallow, fast-flowing waters that turns out to be a life-saver for otters.

One of my first ever memories of fish was witnessing salmon leaping the weir at Wickham Mill, on the River Meon in east Hampshire. Don't get the impression that this is one of those monster waterfalls you see on natural history programmes; the Meon is not a huge river, it is another chalkstream, a bit bigger than the Wallop Brook, the weir created by the outflow of the mill race that ran alongside the building and was no more than two or three yards wide. On tiptoe (I was that young) I'd spend hours leaning over the not-so-very-tall red-brick parapet of the bridge, watching the fish try again and again to negotiate this impediment on their journey upstream. It always seemed to me a mighty endeavour, failure outweighing success nine leaps out of ten, if I'm being generous to the leapers. In the pounded pool below the weir they would circle, dark shadows coming in and out of view whenever the oxygen bubbles momentarily parted. Then from the deepest part of the pool, a yard or so short of the weir, one would accelerate from the water, leaping towards the lip, gliding in the air for a short moment before landing with a smack.

I soon became something of a master at predicting which fish, once airborne (I think with hindsight they were more probably sea trout, which are brown trout running back from the sea), would make it and which wouldn't. It wasn't the speed achieved or the height of the jump that determined success, but rather the re-entry. If they hit the water tail first or head up it was game over, for they were flipped and deposited back in the pool. The same for sideways or for those who fell short of the lip. No, it required a sleek, head-first entry. I had no geometry at that age but now I'd describe it as 45 degrees. But that was not enough in itself to guarantee success; you needed a bit of luck and lots of strength. The luck was a pocket of slack water just upstream to the lip of the weir. I can't possibly attempt to explain or understand the hydrology for the existence of this, but every fish had to hit that exact spot amidst all that pouring water. It was the place that, once submerged, gave a milli-moment of respite where the fish could gather itself before using all that strength to push on through to quieter waters beyond.

It took me some years to understand the full picture; sex and fish biology were still mysteries, but I was constantly drawn back to that pool, staring into the depths, willing a fish to appear but with no real understanding of the purpose of this struggle against the elements. However, as time taught me more about the habits of fish and once I knew the purpose of it all, finding the spawning grounds was the next obvious challenge. Actually, it is not as hard as it sounds. Sometimes we assume fish running upstream to breed have to reach the uppermost headwater of any river. They don't. They simply have to reach a place with sufficient gravel and water for their needs. In fact, headwaters can be dangerous, prone to droughts that may well undo all your good work. So, much to my surprise, with a little advice as to what to look out for, I soon found what I was seeking.

Trout, as with salmon, make no attempt to disguise their redds. They dig them big and bold in the shallows, creating bright, clean patches of gravel that are easy to spot. They are not fiddly little things. Think something the size and shape of a snow shoe for a trout; salmon's redds are somewhat bigger and deeper. In fact, if you see one where all the gravel is gone, dug to the rock-hard sub-surface, you may be pretty certain that it is a salmon. And then there are the fish. Once they have started digging, the clock will be ticking; after all, the urge to dig is ignited by biological changes that will not pause or halt. There is little time to waste and, put bluntly, an unattended redd will soon be taken by others. So find your redd and two fish will not be far away, the male especially bright red in spawning plumage. Bide your time for that moment when they start circling each other before the union, such as it is, takes place. You'll be amazed how close you can approach. I often did. Under the grey skies of winter, with the sun low and the light flat, when the fish only have eyes for each other, a gentle approach from downriver will bring you almost within touching distance. Stand stock still and you'll not be noticed. They will even knock against the rubber of your boots, bouncing off, the glancing blow no apparent impediment to them, though you might be shocked by the impact on your leg – fish are not floppy, they are as strong as steel.

And then suddenly millions of years of evolution will happen right at your feet. It is an amazing thing.

Why do I tell you this? Well, if I as a clumsy teenager was able to make it that close to two fish in their prime, just imagine how easy it is for an otter when the opportunity offers itself. It obviously isn't hard at all; the winter bounty of an egg-ripe trout is a bonus with few equals. Our otter, or any otter for that matter, will tear fast and deep inside the belly of the fish to seek out the two pouches of eggs ahead of any other part of the body. Funny how they know that. Breaking through the hard membrane of the pouch, a long sliver of a thing that would fill your hand, the amber eggs will spill out onto the ground, anywhere from a few hundred to many thousands, depending on the size of the fish. Every single one will be wolfed down, the otter nosing and licking at the grass to make sure none is missed. And only when they are all gone will the otter start on the remainder of the fish.

It is such chance encounters that determine whether our otter will make it to spring. The death of a productive trout is no help to the survival of that particular species, but for Kuschta, Lutran, Willow and Wisp it is quite possibly the end of the beginning.

CHAPTER 12

LUTRAN'S TALE

Estuary otter

It was in the snow of a dark March morning that I last saw Lutran with his mother and sisters. As was typical at that time of year, they were plundering the lake for fresh fish – four dark shadows showing out against the white of the snow as each munched its way through a trout. They were not hard to see and were certainly easy to hear, crunching every piece of the fish from head to tail.

Ravenous they clearly were, hunger ridding them of their natural inhibitions, or at least their normal caution, feeding on the wide-open stretch of grass that surrounds the lake. Oftentimes they'd drag the fish away down to the meadows, or just be alert to what was happening around them, but on that morning not so much. I've long given up the commando-style stealth in my approach; it seems that as long as I am moderately cautious they simply decide whether to tolerate me or not. Some days I'll approach within a few paces, so close that they depart with the utmost reluctance, other times their heads whip round, they fix my progress from however distant and vanish in a trice. Today was the former, as I came close enough to catch the metallic smell of the blood and guts of the dismembering before I backed away to give them some space whilst I went about my daily routine. They have seen it often enough to know I'm no threat, so today my presence is deemed tolerable. I'm grateful for that, though it will forever surprise and delight me that such a thing can be so.

Anyway, back to that routine. The mill has a series of grilles that prevent the lake trout escaping into the river, plus hatches that regulate the flow of water that goes through, under and around the building. The hatches are all very ingeniously designed, dating back to the Dutch engineers of the sixteenth century who created the watercourses that power the mill wheel and sometimes, more importantly, prevent the mill itself flooding. Over the two decades that I have lived here I have experienced more than one 'once in a hundred years flood event', as the authorities like to term what you and I would call extreme flooding. But mercifully never has a drop of water made its way inside the house. Actually, if you want to be sure of owning a house that will never flood, buy an old mill; those once-in-a-century 'events' are so common that any possibility of flooding was designed out of the topography and structure centuries ago.

However, there is one rider to that comforting knowledge – you must maintain and control what those Lowlanders gifted you. Hatches, the wooden or steel gates that hold back or release the river flow, must be raised or lowered according to the levels. Sometimes you are reacting to weather that has already occurred, other times anticipating what is forecast.

More recently, the addition of the lake grilles is an added complication as they soon become clogged by weed, leaves and general river detritus that has to be raked clear twice a day. That was why I was out and about at the pre-dawn of that snowy morning to see the second schism in a family that was gradually breaking apart.

A month shy of his first birthday, Lutran was still far from being the complete male otter. He had much to learn. Should the cards of life fall his way, he would eventually find a new home and sire his own family. But the odds of that happening were far from being in his favour. If it were to happen, and for plenty of his like it won't, there was the best part of another year to navigate. As of now he was about to become a vagrant in his own land, shunned by his mother, disregarded by his siblings and a mortal enemy to any other otter whose path he might cross. For all the dangers and difficulties ahead, he would not be completely bereft of hope. Kuschta had schooled him well. The exuberance of youth was fading. He was an accomplished hunter; catching fish in the lake came easily to him and he was nearly always the first on the bank with a trout. He was young. He was fit. And at this point he was well fed. So was that why Kuschta chose this moment to turn her back on him forever?

I'd like to think she knew it was his time, a sort of maternal acknowledgement that Lutran was fully fledged and ready to leave the nest. But somehow I suspect it was borne more out of her self-interest as both the guardian of the less-developed Willow and Wisp, plus her need eventually to be rid of all three so that she was free to seek out Mion again. The simple fact is that Kuschta's 'territory' was not chosen to sustain what was by now effectively four adults. In the summer and autumn when the family demands were fewer and the living was easier, maybe. But now, at the lowest ebb of the year, someone had to go, and as the four went to leave the lakeside Kuschta put herself between the siblings, letting the sisters go ahead, holding back Lutran behind her. As Willow and Wisp disappeared into the meadow, Kuschta turned to face Lutran, the two halting a few yards apart on the grass.

Strangely, at least to my mind, they did exactly what competing cats do; they groomed in an attempt at feigned indifference, all the while eyeing each other up. Did Lutran know what was happening? I suspect he did. He made a half-hearted attempt to move towards and past his mother, who sprang up with a short bark and hiss, halting him in his tracks. Then Kuschta moved fast towards him, snarling, teeth bared, fur bristling, forcing him to retreat to the edge of the lake. And then, for a moment, nothing happened. They were both frozen, immobile, inches apart, just staring at each other until a whickering 'eek' drifted up from the meadows. This seemed to electrify Kuschta, her head turning first to where the sound had come from and then back to Lutran, before, a decision made, she launched herself at him, wailing like an angry feline. She hit him full force, claws out, ready to fight, but he put up no resistance, tumbling away, off the bank and into the lake. Surfacing, he took one glance at Kuschta, poised on the edge, looking down on him, before he swam for the island without any further hesitation. Kuschta didn't follow him into the water but watched his progress until he had hauled himself out of the water, up the bank and shook himself dry on the island. Apparently satisfied with her work, she turned away and without a backward glance was gone, her progress down the brook marked by the eeking calls of mother and daughters seeking reunion.

No doubt Lutran heard all this, but he made no attempt to re-join them. For a little while he padded around the island, eventually settling down in a clump of nettles at the base of the ash tree. Much later in the day, under cover of some rain, I saw him swim across the lake, heard him plop into the brook as he headed for the meadows. I'd like to think his next stop was the shelter of the willow holt I'd built, but I'll never really know. That was the last time I ever saw Lutran at the mill.

Young male otters such as Lutran have some simple rules to follow once they have split from the family group; keep well fed, keep out of trouble and keep going until you find a new home. There is no back-up or alternative lifestyle. Returning to mother for succour or seeking the protection of the father you have never met is not going to happen. Nor is there a collective of adolescent otters who look out for each other. You are always competing for the two things you most need: food and territory, and you do that on your own, alone in the world, often in places you have never been and with foe you rarely see. That was the task that greeted Lutran as he emerged on that rainy dusk evening, alone for the first time in his life.

Heading down the brook was a familiar route for him, with recognisable landmarks and the comforting scent of spraints that told him the others had passed this way not so long since. But he made no attempt to rush on or catch up, as it were. Whatever part of the otter brain that registers these things told him it was better to know where they were than to be with them. Just short of the deep pool below the road bridge, he paused for a while, sheltering under the arch. From time to time the tunnel reverberated as a car passed overhead, but the rumbling disturbed Lutran not at all. This was a regular resting spot and the familiar sound was a comfort. But rest was not what he needed; rather, it was food, and the pool had a particular characteristic defined by the original construction of the bridge that would help him claim his first solo meal.

Long ago, when the bridge was built, the river bed was lined with heavy slabs of stone; I'm no civil engineer but I guess the purpose of this was to prevent the foundations being eroded. The net result is that the river is constricted, speeding up as it passes between the brick uprights, tumbling out the other side, pouring over the lip of the slabs that extend a few yards beyond the bridge. Over the centuries water has dug the deep pool, the back current eroding a dark cavity beneath the slabs themselves. It was there that Lutran was headed. If you have ever swum under a waterfall you'll know that once you get beyond the pounding curtain of water there is a calm place on the other side. Well, in its own way the cavity was that place, and as Lutran nuzzled along the sides and floor he was searching out others who found this sanctuary equally appealing. It was a place that took little effort to live in as it was forever replenished with new food being washed in. He was, of course, on the hunt for crayfish, which were equally on the hunt for snails, pea mussels, nymphs and just about anything in the river you might reasonably consider edible. But it wasn't the wisest spot for crayfish to ply their trade – there were no loose stones, layers of mud or bits of vegetation under which to hide or flee for cover. The cavity was washed shiny smooth, something of a crayfish death trap once Lutran locked on to his prey. Four times he dived and four times he resurfaced with a new victim, perching on the ledge under the bridge where he chewed each up in turn. After an hour he was done, but before he went on his way he deposited his mark. For the first time in his life he was claiming a place as his own. In truth, this was on the side of vainglorious; Lutran was elevating his achievement beyond his current status. Any following adult would dismiss the scent of his spraint for what it was – the mark of a displaced adolescent. But no matter, Lutran was at least in the game. For the remainder of the night he kept on going downstream, passing through places he knew well, but the crayfish were to be the last food on that particular evening. Whether the places he knew had already been fished out by others or his skills were simply inadequate to the task, only time would tell. But for now, as dawn came, he was more tired than hungry, sniffing out an empty holt where he curled up for sleep.

Lutran woke at midday with a gnawing hunger but with no way to satisfy it until dark. The holt provided refuge but no food beyond a few insects and earthworms and he wasn't quite that desperate yet. Outside the rain fell. Inside he kept himself dry and warm, waiting for the night, conserving every ounce of energy. Though he did not know it, he was travelling in reverse the self-same route by which his mother had arrived at the Wallop Brook a year earlier. Like she had done then, he was doing now, venturing into unknown territory, exploring new lands, uncertain about what was to be found behind each new bend in the river. By dint of his circumstances Lutran can't move fast; every spraint must be examined, every holt or couch inspected. He is forever looking for that territory where there is no evidence of fellow otters, or at least where the evidence is so old that he might relax. But this, at least for now, never happens. Over the next few days he manages to evade direct contact with any others and even at some point left Kuschta, Willow and Wisp behind him. Eventually he came to the confluence where the Wallop

Brook meets the larger River Test, and for no other reason than that the current took him that way, he headed downstream in the direction of the sea. Albeit accidental, it would turn out to be a fateful decision.

Moving down the main river was a very different affair for Lutran than down the brook. In the past he'd been used to a stream where he could almost jump its entire width in places, where progress was as much about scrabbling over gravel bars or paddling through the shallows as it was about outright swimming. Now life was on a far bigger scale, with a river faster, wider and deeper than anything he had ever experienced. Suddenly there were banks so high he could not possibly scale them. Pools so deep he'd have to surface for breath on numerous occasions before they were fully explored. Structures across the river the like of which he had never seen before. And people – the noise and activity of humanity was on a scale that he was experiencing for the first time. But remarkably he took much of that in his stride, for his mind was occupied by a single thought: food.

The last few days had seen meagre pickings. The big river required a whole set of new hunting skills that Lutran was yet to learn. He felt safe carried along downstream with the current, but it was no way to find food. Often his whiskers would pick up vibrations, telling of a fish close by, but in that volume of water, many hundreds of times greater than anything he had ever coped with before, the tell-tale pulses usually faded into the distance and another hour of hunger ticked by. From time to time he moved along the bank, but away from the water the landscape lacked the immediacy of the Wallop Valley. Here it is flatter, more open where the pasture is grazed tight by sheep or manicured for fishermen. There were no Badlands to which he could retreat, eking out a meal from wild, unkempt wetlands. Sometimes he chanced upon food; rotten apples that had lain on the ground since autumn, corn from an overturned pheasant feeder. It was all something to fill his stomach but not enough to sustain him for long.

On the fifth night salvation seemed to come his way, the unmistakable smell of fish catching his nostrils from somewhere away in the distance. Following a little side stream off the main river, suddenly things seemed more promising. Lutran scrambled out of the water, padding along the verge of a gravel track, the scent guiding him, but he knew not where. In fact he didn't really care; the thought of a fish was enough to crowd out any thoughts of danger or curiosity. The food was the thing. As he rounded the corner of a large red-brick building, ahead of him were row upon row of stews, glistening under the orange glow of the overhead lights that illuminated the fish farm. Stews are the homes for captive-reared trout, rectangular ponds that are generally about the length and width of a cricket strip, fed by the pure, fast-flowing chalkstream water that flows in one end and out the other. For the trout, usually North American rainbows that grow faster and better than our native brown trout in conditions like this, it is a bit like living life in one of those exercise pools where you constantly swim against the current but resolutely remain in the same place. It sounds a bit harsh, but really it isn't. The fish thrive in these imitation rivers which are probably reminiscent of the fast streams of the Rocky Mountains from where the original brood stock was imported back in the mid 1800s.

Lutran knew or cared nothing of that, lured only by the tempting odour the stews emitted. If you ever visit a fish farm you'll notice the same smell; it is not unpleasant, it is a sort of dry mustiness that makes you sniff two or three times in the expectation of something worse, but it soon desiccates the senses to the point where you notice it no longer. What you will notice, however, as Lutran soon did, is that every stew is protected by netting or steel grilles – in this case the latter. As he sniffed around the edges, seeking a way to reach the swimming trout, a whirring sound, followed by a rapid cascade of plopping then an explosion of noise, halted him in his tracks. The fish in the stew were suddenly in a frenzy, cavorting at the surface, their silver bodies catching the reflection of the orange light as they swirled around like so many writhing snakes, in fierce competition for Lutran knew not what. Then suddenly they stopped, disappearing back under the dark water, just a few cruising dorsal fins and backs showing above the surface as the commotion of the waves gently subsided, leaving only the regular stream of the water in one end and out the other.

As he made his way from one stew to the next, Lutran was constantly thwarted by the steel protection, the fish temptingly close but never in any sort of danger, at least not from him. For time to time as he snuffled around, frustrated at his inability to get into the water and at the fish, he'd come across a little brown pellet, about the size of a pea. The first few he sniffed at suspiciously, but they smelt, well, of fish, which was invitation enough so he took to eating them. They tasted good. They would do. After all, they were the pellets that fed the trout (that whirring sound was the automatic feeders that dispense pellets hourly) made up of 50 per cent soya and 50 per cent fish. Packed with oils and all kinds of stuff to help fish grow, the pellets were far from bad for Lutran; it was a pure protein hit. I doubt he immediately appreciated this, but he soon had the taste for them, nosing into every crevice and seeking out the stray ones as he padded along the paths between the stews.

It wasn't just pellets; remnants of dead fish were also there to be found and Lutran found them. Fish farms are like that; mortality is a way of life. When you have tens of thousands of live fish a goodly few die every day through natural attrition. That's how nature works. Occasionally that can run to hundreds of deaths if conditions are extreme – extended periods of hot weather can deplete the oxygen level to the point where the fish literally suffocate. At the other end of the scale huge volumes of dirty flood water can be similarly depleted. Fortunately, most fish farms have emergency 'bubblers' that kick in as such crises approach, pumping aerated water into the stews to keep the fish alive until the worst has passed. Regardless of all that, Lutran was at last finding plenty to fill his stomach. Most of it wasn't what you might call fresh: a small, almost dried fish that had fallen out of a collection bin and lodged in some long grass, a squashed trout pressed into the gravel that had been caught under the wheel of a tractor or some such, plus tails, entrails and general bits of trout that probably only a desperate otter would seek out. Actually, that isn't exactly true...

The other great scavenger around the farm that night was the common brown rat, *Rattus norvegicus*. These are far from being the grey creatures you might see in a city sewer or disappearing down a drain hole. If there were a *Rattus* beauty pageant, our fish-farm inhabitants would win best in show every time. They look in positively rude health, plump with chestnut brown hair that glistens. Even if you don't like rats (and most people don't), it is hard not to admire them for being so buff. And there is no reason why they should not thrive; fish farms were almost invented for them, with food in abundance. The places Lutran can't reach, a rat certainly can. They squeeze through tiny gaps and are remarkably good swimmers; water holds no fear for them. One stew might contain fifty thousand fingerlings, an easy dinner every time for these resourceful creatures. And like Lutran they soon understand the value of a good fish pellet.

As he searched out food, Lutran both smelt and saw the rats; they were something largely alien to him. Brown rats for the most part inhabit the places people inhabit, the polar opposite of the empty wetlands, woodlands and meadows Lutran had been brought up in. Water rats, aka water voles, he'd seen pretty well every day of his life, but as they are largely herbivorous their respective food chains don't conflict, so otters and voles generally just ignore each other. But the rats were different. Lutran could sense they were wary of him, melting into the darkness as he moved around. This he could not quite figure out, but when he stumbled upon a rat chewing on a fish head, coming eyeball to eyeball, its reaction gave him a first notion that otters are the rulers not the ruled. There was no hesitation on behalf of the rat; it dropped the head and fled, leaving another small meal for Lutran that he gratefully accepted.

Emboldened with his new-found place in the world, Lutran took to exploring the farm. He sniffed at the pungent aroma of the diesel water pumps and scrambled through the large plastic pipes that are used to transfer the fish from one stew to the next. Guided by a now familiar smell, he even found the pellet store, hopefully clawing at the rat-proofed door before giving up a hopeless task. For a while he even curled up on a pile of soft nylon nets to sleep. But never once did he manage to get close to a live fish, soon ignoring the siren call of those whirring feeders and the splashing that followed.

He'd try again tomorrow. However, as the dark night turned to a grey, drizzly dawn he had succeeded in sating both his curiosity and his hunger, settling down to sleep in a dry storm drain.

He woke to sounds. In fact his entire day was to be punctuated by noise; his cosy drain was smack in the middle of the farm. Tractors trundled above. The pumps were fired up, the diesel fumes drifted his way, as did the clanking noise of machinery, pipes being coupled then uncoupled. People stopped nearby to talk, never once suspecting they were within a yard or two of an otter. Not that Lutran gave them any indication of his presence, pressing himself into the darkest recesses of his makeshift holt. His life so far gave him no particular reason to fear people, but centuries of evolution have taught otters to be wary of their one natural predator. But eventually the sounds faded, darkness came and the farm was once again deserted, bathed in orange light. That night he made the most of his new home. True, he never found a chink in the armour of the stews but he did find the sacrifice pond, upgrading his diet of dead fish and pellets with something more familiar, hauling himself content and replete back to the drain well ahead of dawn for another day of sleep.

On what was to be his third and last night at the farm, Lutran stood motionless at the mouth of the culvert, only his head visible to the outside world. His nose and ears worked the night air for the tiniest indication that he might not be alone. But all seemed well so he slid out, resuming his restless exploration of the farm. Fresh spillages of pellets were soon found, and the nets, still wet from trawling stews, had a few tasty fish parts lodged in them. Intent on his scavenging, Lutran failed to take any account of his surroundings. To be fair, he was still young and immature, but that failure left him vulnerable to attack, and, sure enough, it came. No trout farm will ever be an empty territory; the pickings are too rich to be ignored. From a far corner Old Dog made his silent move, a scarred and seasoned otter who had rights to the farm, back from travelling the outer bounds of his territory. Moving in the shadows, he had Lutran in his sights, yet another claimant to be dispatched. With five years and five pounds advantage, Old Dog had no reason to fear Lutran. He had done this many times before, waiting until he was within two bounds before launching himself at an unsuspecting Lutran with a wicked, whickering battle cry.

Lutran's first move was to flee, but Old Dog was having none of it. A pound of flesh was demanded as he leapt on Lutran's retreating back, biting at his neck. Lutran whipped round, rising onto his hind legs, the two pawing and clawing at each other, heads jagging forward to inflict sharp bites to the head, face and ears. Like kick boxers, they use every limb to inflict damage or give leverage. But, unlike human combatants, there is no respite. No clinging together to catch breath. No bell to sound the end of a round. It is continual conflict as they roll and throw each other over, the whole time biting and clawing, seeking out vulnerable, soft areas of flesh for maximum injury.

It was far from being a fair fight. Aside from his size, Old Dog had dozens of conflicts under his belt, plus the element of surprise and the fierce desire to defend his territory. Soon he had Lutran cowed, standing over him, jabbing his front claws into Lutran's face, who squirmed and turned, retreating from the onslaught the best he could until he saw a gap, running headlong towards the river, launching himself into the water in the hope that his attacker would not follow. As he dived into the river the cold water bit into his many wounds, and for the first time Lutran felt not just fear, but pain, too, as he swam away down the stream that had originally brought him to the farm. Naturally, Old Dog did not follow. He had more than made his point. Otter justice had been meted out. For a short while he watched as the tell-tale bubbles marked the retreating path of the vanquished, before he sauntered back to the nets to resume the job Lutran had started.

That night, and the following day and night, were the low points in Lutran's short life as he found refuge in an abandoned rabbit warren in a small copse beside the main river. Nothing, not even the rough play with his sisters or the final showdown with Kuschta, had prepared him for the sudden all-out aggression of the old dog otter. Some of his wounds were deep and long, his pelt cut open, exposing the pink and white flesh beneath, which he licked and kneaded as best he could, all the while too stiff, bruised and scared to venture outside. But for the most part he slept, tucking himself into a tight ball, letting the passage of time heal his body. By the third night any residual fear was seeping away, replaced by hunger. However much it hurt to move, Lutran had no option but to leave the safe comfort of the warren to go in search of the food he so badly needed. The text books say of otters at Lutran's stage of life that they are sub-adults who live sub-optimally. It really isn't a very uplifting description. In a way they deserve better. They are, after all, struggling against great adversity to make it through to the next level. But though it is a bleak assessment, it is very true. Lutran was a long way from being an adult and he was living life on the fringes, trying to survive in marginal territories with inferior skills. The fish farm had been a lucky break with an unfortunate ending. He now needed something more to go his way.

Heading out on the third night after the fight, Lutran continued to follow the River Test downstream; he knew enough to realise that the places he had already been were not places to dwell. Soon the countryside started to run out, with houses, roads, buildings and all the other manifestations of the human world increasingly impeding his path. He was on the outskirts of Romsey, the largest town on the banks of the Test, just ten miles from the sea. Against his will he was forced back into the water; his partially healed wounds hurt so much more when wet. But now the river had become a channel, bounded by concrete and steel walls built high to protect the good burghers of the town from flooding.

The winter flow helped Lutran made good progress, but with the sheer sides there were few places to stop and rest, not that he felt much inclination to do so. Often the river was lit by the spillage from street lights, so Lutran took to hugging the shadows, slipping by unnoticed as he traced his way under the tenth-century walls of the Abbey. Like at the fish farm, but now on a far greater scale, he heard the noise of traffic and people, so when the river went underground or through a tunnel the silence and darkness were a blessed relief. But progress did not equal food. The riverscape was simply too unfamiliar, confounding his attempts to find anything to eat. It was fair to say he was getting desperate, weakening with hunger, but as the town began to peter out the river became less of a channel; trees replaced houses, steel shuttering gave way to soft earth banks, the glow of the town was now behind him and darkness was ahead.

Somewhere out there, ahead in that darkness, Lutran heard the pounding of water, a roar as it tumbled off a steep drop. He could feel the pace of the river increase beneath him but he was too tired to swim against or across the current as the river narrowed, funnelling the water through some unknown, narrow structure ahead. The closer it came the faster he was carried. He could do nothing to slow his pace or divert his course beyond keeping his head pointing downstream, his nose above water and his tail stretched out behind him, acting as his rudder. Then in the final moment as he approached the tumbling drop-off, the current from the left and the right slammed together, pincering him in the middle as it gripped him, spitting him through the opening, out to whatever was on the other side. For a fraction of a second Lutran was airborne, weightless until gravity took over, tumbling the powerless otter into the water, which again held him as the vortex drew him deep, deep down. Pushed ever deeper, Lutran could do nothing to counteract the force of the water, so he just let his body go with the motion, randomly spun like an out-of-control acrobat amongst the fizzing, champagne-bubbled water. Gradually the current slowed, the fizz dissipating, a little vision coming back as the water pressed down less on him. Still deep and disoriented, Lutran righted himself, swimming with the flow, letting his natural buoyancy slowly take him towards the surface. With a cough and stran-gled bark he broke into the fresh air as the river washed him onto a gravel spit. Climbing up onto the dry promontory, he shook himself down and turned to look back. He had been carried fifty yards downriver from the salmon leap at Sadlers Mill that had sucked him in and spat him out.

For a while Lutran groomed then rested, but none of that did more than momentarily relieve his hunger and exhaus-tion. He needed food more than anything, but going back into the water was a miserable choice so he reverted to his younger self, sniffing along the shore, gently turning over stones with his claws, licking up the nymphs, snails and worms that he exposed. As he scavenged amongst the river's edge where the water lapped, the moonlight caught the peaks and troughs of the gentle waves. Then just ahead of him Lutran spotted a familiar outline, the back and dorsal fin of a silver fish breaking through the surface of the water, immobile except for the motion of the waves.

I say familiar, but it was and it wasn't. Yes, it was clearly a fish, but the fin suggested one bigger than any he had encountered in the past. Secondly, it wasn't behaving as a fish should. No fish in Lutran's past experience lay inert in the shallows, vulnerable to all sorts of predators, including him. Had it been smaller, I suspect Lutran would not have thought twice, rushing across to the gravel to grab the fish, which in all likelihood had heard him coming, escaping easily with a single flick of the tail. But the size and position gave him pause for thought. He gingerly stepped back a few paces, careful to avoid any clack of the gravel before lowering his body into the water, quietly pushing out into the main current, swimming in a semicircle that, if he had planned it right, would bring him around to attack the fish broadside from the cover of deeper water.

It is no surprise that Lutran had never encountered a fish of this size; it was a salmon, far bigger than anything that would ever have made its way up the Wallop Brook. The reason for its position in the soft, easy shallows was that it was a kelt, an exhausted and depleted post-spawning fish that was making its way back to the ocean. Far back in the autumn, it had swum from the ocean into the River Test, gradually making its way upriver, negotiating many obstacles, including the salmon leap, until it eventually reached the soft gravel spawning beds somewhere near the headwaters. But in all that time, despite the huge effort required to make that journey and all the rigours that spawning entails, it had never once eaten anything. For months it had been living off reserves built up whilst out at sea. It was, in so many respects, a fish that was a shadow of its former self, its three-foot frame clearly emaciated. The rounded, deep belly had gone. The tail and head looked disproportionately large. The fins and tail were ragged. The red filaments of the gills were crawling with maggots. If you'd have picked up this fish fresh from the ocean you'd have felt an animal as strong as steel, whipping in your hands. Today it would be flaccid and inert. But, for all that, you would still recognise the sleek shape of an Atlantic salmon, with bright sides glinting like hammered silver with a grey gunmetal back. All it would take would be a single day of swimming, perhaps less, to reach the tidal estuary and to have the chance to feed again before heading back to the Greenland coast. Plenty do survive their time in the river; some even make the journey across the Atlantic and back in successive years, but for this particular fish it was not to be.

For all his inexperience, once Lutran had set himself for the attack, the salmon stood no chance. Lutran swam at full speed from the open water directly at the head of the salmon, biting down somewhere around the back of the neck. His momentum was enough to carry him and the fish up the gravel bank, the two somersaulting as they went. As the fish flapped, Lutran shook it violently, the salmon going limp. For a while Lutran lay on the fish, uncertain of his victory, clawing into the sides for a better hold on his capture. His teeth were still deep in the flesh, whilst he wheezed and gasped through his nose to get his breath back. Then he stood up, lifting the fish with him, shook it again and, finally satisfied that it was dead, released his hold, letting it fall at his feet.

An older, wiser Lutran might have taken stock of his surroundings before devouring the fish. Some judicious grooming perhaps, whilst taking in all that was around. Or maybe drag the fish to somewhere less visible. But not young Lutran. He was too hungry for any of that, but he knew enough to start at the underbelly, tearing it open with his sharp incisors before pushing his head inside the still-warm body to seek out the blood-rich liver, heart and kidneys. Eventually, when he was done with that, he did take some time to groom, washing the sticky blood from his head and neck before resuming his feast. Lutran was a bit like one of those diners you see at an eating contest, bolting the food at the start, mouthful after mouthful disappearing at speed. But gluttony can only take you so far. There comes a point where each additional bite takes extra effort, the food chewed over and over again to overcome the gag reflex. To be fair to Lutran, that point was an hour in the coming. He was done, too sated to eat any more, although the salmon, now a raggedy, torn version of its beautiful former self, was still only half eaten. Finally Lutran did take some time to look around, padding the length of the promontory, crawling under a morass of roots that had created a curtain along the bank, dark, dry and dusty on the hidden side. Sniffing it out, he knew it was good enough for now, so he dragged the salmon inside, curling up for sleep.

The following night Lutran did nothing but eat, snooze and groom, finishing off the remainder of the salmon. Otters are like that. There is no real forward planning. You might think that an hour or two of hunting to double up the store of food might be advisable, but apparently not. I guess that is the thing about being an apex predator – you really don't feel the need to bother. On the third night, the fish all eaten,

Lutran headed out from his temporary holt to explore the huge pool below the salmon leap. He swam up to the cascade through which he had tumbled, careful to circumnavigate the pounding waterfall that flows out of one of the hatches that releases the water around the side of Sadlers Mill. To the left, in the smooth stone wall that creates the top arc of the pool, is a square entrance that gives way to a dark tunnel, a far gentler current flowing from it. Lutran paddled slowly inside. It felt damper than outside, a musty smell reaching his nose whilst the slap of the gentle waves echoed across the chamber, the vaulted brick ceiling towering above him. Ahead, the giant mill wheel stood idle, the lower blades half in, half out of the water.

The mill wheel housing was where Lutran felt safe. The darkness was familiar. Even the musty smell felt like home, reminiscent of the earthy holt in Beech Wood. Clambering onto the blades, he spied a wooden walkway above him, and by stretching himself to his full height he grasped the edge of the rough oak boards with his front claws, his lower legs scrabbling for purchase on the wooden upright to haul himself up. The narrow walkway, just wide enough for the millwright to make his way around, took in three sides of the housing, was part rotten and showed little sign of human use. Likewise, it was a while since this particular wheel had spun; the cast-iron bearings on which the axle of the wheel had sat had long since rusted together. Sniffing around, Lutran found nothing he didn't expect; the droppings of birds, rats and bats festooned the boards. They liked the refuge as much as he did. All very familiar. All very good.

Hopping over the axle to check out the other two sides, his nose caught the scent of otter and within five paces he was at the corner, a pile of spraints before him. In an instant he knew he had to leave; the latest was so fresh that whoever had been here had probably passed through sometime whilst he was out on the promontory. His olfactory senses gave him all the data in one quick hit: a male, older than him, and this was a taken territory. The spraint heap dated back months, possibly years. Lutran didn't hesitate; the memory of the fish farm was enough to trigger his flight as he scurried back down the walkway, over the axle, and dropped into the mill race below. He didn't surface until he had reached the middle of the pool. He had no idea whether the otter was close at hand, but he was taking no chances and figured mid-river to be as safe as anywhere. At least he had plenty of options for flight. Paddling water for a few moments, he scoured the tall wall at the head of the pool. There seemed to be no easy way back upstream, so, fearful of hanging around any longer, he turned with the current, swimming he knew not where, as befitted the stateless otter he had become.

As it so happens, Lutran's fear made for a good decision. As he swam through the night, though he didn't know it then, he was getting closer to a place he would be able to call his own. As dawn approached, the landscape became very different as he went from river to estuary, passing under the last bridge before the ocean. Suddenly he felt a little more buoyant on the brackish water; swimming was just that bit easier as he rode higher in the water, though the saltiness was something different. No longer were the banks green and mostly linear. Suddenly it was all a jumble, with inlets, bays, brown reed beds and mud flats. He felt the pull of the outgoing tide taking him along. At one point he tried to walk to the shore, but he sunk to his chest with each step, soon giving up the effort as too muddy and tiring.

Eventually the tide diminished and then finally went slack,

encouraging Lutran to come to a halt. After a few false starts he found a sand bar that led to a little tussock-covered island. Around him Southampton Water was filled with industrial life. The red lights on top of the cranes that served the container dock speckled the skyline. The glowing tongue of the gas flame from the tall oil refinery chimney veered this way and that with the wind. Directly above him the thick cables of the power lines, suspended from the steel pylons, hummed with electricity as they travelled north. Occasionally the horn of a ship echoed around the marshes.

But none of this bothered Lutran one jot; he had no reason to fear any of it, so he ignored it all, for as far as he could tell he was in a place where no other otters lived. As he wriggled and rolled in the rough tussocky grass to rid his fur of the itchy, drying salt, his sole concerns were for sleep now and food later. Both would come very soon, as he was about to become an estuary otter, but maybe that is a story for another day.

CHAPTER 13

COMING OF AGE

The duckling decimation

May was very different to March for all of us. The long slog from winter to spring was over, and summer lay just around the corner. The valley had gone from denuded stark to verdant green. The birds sang, the brook burbled. Around every corner new life was emerging. It was a good time to be an otter.

In truth, since the schism on that snowy dawn in March I hadn't seen much of the family around the mill, largely, I think, because they had pretty well stripped the lake bare by that point. Fish-catching will always be the hardest skill for any otter to acquire. It takes guile, patience, agility, strength, speed and experience, all qualities that Willow and Wisp had in part but could not yet always put together. Mothers can only teach so much; the remainder comes from ever-evolving trial and error. Very soon their survival will pivot on less error and more success, but until I re-stocked the lake for the new fishing season the few fish left in an otherwise empty pond were simply beyond their hunting abilities. That said, it didn't deter Kuschta, who still made the occasional solo visit, appearing at all hours gliding around the lake in search of a really good meal.

By now only the fittest and most cunning of the trout had survived, but there were not many of them; my daily feeding only brought a handful to the surface, and instead of a roiling, boiling mass of greedy mouths churning the water to foam, the few survivors calmly cruised on the surface, glugging down the pellets in no apparent hurry. I'm sure Kuschta made a point of arriving alone because of this, knowing that her two inexperienced, clumsy daughters would have no chance, spooking every fish so that even she wouldn't be able to catch one. It also seems pretty clear that, as the one-year anniversary of their birth approached, maternal patience, if not exactly wearing thin, was at least in its final denouement.

But for all that, on the few times I did see them together they still seemed to be that contented, cohesive group that had a complete belief that the world existed for their benefit, where very little impinged upon what they did or how they organised their daily existence. Otters are in some ways a total contradiction. On the one side of life they are solitary creatures, who, if not exactly shunning their fellows, do everything to avoid all but necessary contact, which is essentially borne out of either conflict over territory or those handful of days for courtship. If you put that in some sort of statistical context, an adult otter, male or female, will spend less than one half of one per cent of its entire life in the company of another adult otter. There cannot be many other creatures in the animal kingdom who are so determinedly single. On the other side of life, when forced together as a group, that solitary gene just vanishes. In captivity large groups, even including mature males, will happily co-exist for years on end. Visit the otter sanctuary at any zoo or wild-life park and it will be hard to be anything other than captivated by the life of otters as they play, dive, stretch out in the sun and indulge, at the very worst, in a few petty squabbles. You could be forgiven for thinking that was how otters were all the time. Likewise, in the year or so that I watched Kuschta raising her family, aside from those painful moments of separation I saw nothing but a mother who took great care of her pups, who in turn stuck to her like glue, absorbing her every move.

As has been said, there is no specific breeding season for European otters; litters are born in every month of the year. But if I were to come back as an otter I'd choose to come of age when the living is easy. May or June wouldn't be such a bad choice – the fates had smiled on Willow and Wisp. I must admit, though, that it was getting harder by the day to distin-guish the two from their mother, as they were by now all of very similar size. I certainly couldn't tell who was who when they were in the water, so I generally relied on their actions to pick out Kuschta. For instance, when they arrived it was she who led the pack of three; her frame was just that much fuller than the other two, and once I had that fixed in my mind I was easily able to tell them apart. Kuschta's fur is also a shade darker in colour, but at night and especially when wet, that, not very surprisingly, doesn't help much. Actually, when all three were wet, it was much easier to notice her heavier musculature. And the pups, though I am not sure that, at a year old, they should really still be called pups, generally act like naughty teenagers, competing behind Kuschta for second in line, joshing with each other as they had done since the very first time I had seen them. That said, they were now more inclined to peel off to pursue some separate prey, so sometimes those reassuring 'eeks' were quite distant, triangulating their positions many hundreds of yards apart until eventually you'd hear the trio come together before they headed off into the night. But otters are not alone in enjoying this time of year, and summer becomes a virtual 24-hour bucolic metropolis.

I've spent a great deal of my life on rivers. A large slice of that has been practical, going about my daily tasks, whilst the rest has been simply for the pleasure of it – fishing, ambling along the bankside, or using any excuse imaginable simply to steal time by the water. Rivers soon became something of an addiction; there the hours fly by, the bigger world fades away and the smallest incidents bring the greatest delight. But I am ashamed to admit that for years I knew nothing of otters. In fact, I assumed they were all but extinct on the chalkstreams of southern England. That seemed to be the general consensus, and it wasn't until I was in my late thirties that I saw what was, unequivocally, an otter. Of all the times and situations on which this could have happened, it was one of the more unlikely scenarios. There was none of that dead-of-night stuff or covert observation after hours of waiting. It happened entirely by chance. There was I, standing mid-river, rod in hand, gently wading upstream on a bright, sunny summer morning. All was right in the world on this pretty headwater of the River Itchen, twenty miles to the east of where I live today. As I scoured the smooth water ahead, picking out the hatching insects and looking for a rising trout, a bow wave appeared from nowhere, and the surface was then broken by a furry head. At thirty or forty yards out, I had absolutely no idea what was heading towards me, so as I stood still it just kept on coming, coming, coming, until the 'thing' was so close that I could have touched it on the nose with my rod. I truly believe that right up to that moment the otter had no idea what I was and certainly didn't consider me a threat – the route downriver never deviated nor the pace change. I'm guessing I was simply some unfamiliar obstruction eventually to be swum around, and, in any case, with their famously poor vision, I was probably nothing more than a blur until seen up close.

But close up I was clearly something of an oddity, as the upturned whiskery face looked me in the eye, did some instant calculations, decided I was best avoided and without pause to its swimming unhurriedly executed the most perfect 90-degree turn, disappearing under the water. A moment later he, or it could have been a she, was at the bank, and, with that fluid movement that is now so familiar to me, was up, out and off into the meadow, a few twitching tops to the reeds and sedge grasses marking the route away from the river. What stays with me to this day from that first ever sighting, which can't have been more than ten seconds from start to finish, was the sheer size of an otter, plus that sense that this was an incredibly strong and agile creature. In my entire life I had never seen anything that big in a river, and the way that it so effortlessly turned and vanished is burnt into my brain.

And what did I do? Well, it is pathetic to relate that I went in pursuit of the otter, dragging myself from the river with considerably less grace than it. What I really expected to achieve, I had no idea. The chances of me ever keeping pace, with or without waders, across the boggy terrain of the meadows were precisely nil. For a few dozen yards I squelched between the clumps of sedges, skirted the most obviously wet ground and pushed through the tall reeds until I finally gave up. Quite frankly, the otter could have been within a yard of me and I'd have had no idea. So I plonked myself down on one of the sedge grass heads as a comfortable seat, as I gasped for breath and scanned the meadow for some tell-tale sign of movement. Of course I saw precisely nothing; the otter was either long gone or curled up on a comfortable couch.

Even after that first sighting I'm not sure I immediately thought too much more about otters, putting it down as an aberration, a once-in-a-lifetime moment. After all, at that point (we are talking early 1990s, the nadir of the population) I knew plenty of river keepers who'd never seen an otter. But those fleeting seconds planted a seed. No longer did I instantly assume that reports of stoats or mink were always as we thought. I know that, when put side by side, they are entirely different animals both in size and colour, but a momentary glance seen out of the corner of an eye – well, who knows? After that first sighting, even though I'd seen it plain as day, I'd rushed home to check a text book to be absolutely sure of what I'd seen. From then onwards those partially eaten fish I found on the bank were examined a bit more closely; clearly not every one was a heron or mink victim. The pink fragments of crayfish shell or the occasional pincer claw discarded in the grass suddenly made more sense. And for years I'd paid no attention to the animal paths that tracked along the river bank occasionally deviating into the river, dismissing them as deer or rabbit trails. But closer examination of those muddy slides and paw marks told of something else. Now I'd instantly recognise them for what they are – otter highways.

There are times today, with the otter population growing with each passing year, when tracking otters is not as hard as you might think – I'd highly recommend heading out after a fresh fall of snow. Cold and icy? That holds no fear for otters. They go about their business as ever. Their tracks, the patches of blood and the dark stains of spraints will be plain for you to see. Such will be the state of the battered-down snow that I guarantee you'll wonder if it was a herd of otters that had passed your way. You might not see them at that particular dawn, but you'll learn enough to know they dwell in your neck of the woods. Give it time, with or without snow, and you'll soon piece together the mores and habits of your particular otters. They are truly creatures of habit; you will get to know the routes they follow, the places they eat, the times they come and go, until, quite suddenly, and probably when you least expect it, an otter will become part of your life. You'll never achieve closeness in the way that people sometimes do with foxes, having them eat out of their hand, or with badgers arriving nightly for a feed of apples on the front lawn, but, with a measure of standoffishness on your part, otters will offer you a chink of light that will shine onto their lives.

The departure of Lutran came and went almost unnoticed by Willow and Wisp. That is the thing about animals; they don't mourn the passing of a sibling. It is, in the mantra of a motivational trainer, not a problem but an opportunity. The nightly hunting grounds were now split three ways rather than four. Vying for the attention of Kuschta suddenly became that much easier, too. Life in general was more on their terms. It wasn't that their brother had been a bully; otter families are too collegiate to tolerate that, but his extra size and greater rumbustiousness had always placed them firmly behind him. As for Kuschta, well, the move from the original four to the current two was all part of a harsh evolution. One had had to die, one had had to be driven away, and very shortly the final two would find themselves abandoned. But for now, on a clear July night, as she curled up on a reed couch in The Badlands, she was content to let her daughters explore alone. The more they became used to that, the easier the separation would be.

Willow and Wisp were companionable explorers; the world was still a place of fascination and discovery and they were emboldened by paths and territory that were comfortingly familiar, with landmarks to guide them back home. They were of an age, at fourteen months, when their in-built guidance system was starting to kick in. Otters clearly have an innate ability to memorise a vast network of paths, sometimes called runways, that will crisscross dozens if not hundreds of square miles. With all that sprainting, it is tempting to think that navigation is all about smell, but it seems something else is at work. Nobody has, as yet, adequately explained the 'something else', but it certainly exists, as scenting will only take you so far when water is involved.

It doesn't apply to our Wallop Valley otters, but in regions where there are large lakes or lochs otters will cross them by swimming the identical route time after time. If those same bits of water freeze over they'll cross the ice along precisely the same path. Since, in such instances as this, they cannot be guided by smell, they must be guided by landmarks or some in-built geolocationary sense. Incidentally, otters have perfected a rather charming way of traversing ice, which entails running at speed and then sliding as far as the momentum will take them before repeating the same actions over and over. It certainly looks exuberant, tapping into that vein of thought that otters are fun-loving creatures, but my suspicion is that it is simply the most expedient way to traverse a surface that otters aren't very well designed for.

There is also plenty of evidence, some of it with an unfortunate conclusion, that otter runways are historic, the same routes having been used for hundreds of years. The journals of otter hunts certainly support this; generations of huntsmen combed familiar places time and time again and had no problem picking up the scent if the pursued tried to throw off the hounds by fleeing alternately on water and land. In Norfolk otters still travel the path of rivers drained centuries ago, even though the canals that replaced them flow nearby, and more recently it has been recorded, such is their determination to keep to the same heading, that many have been killed where new roads have been built across ancient runways. So, if you ever chance upon a warning road sign that pictures an otter, do keep an eye out, for it will be one of those places.

Whether Willow and Wisp padded along that river bank that night guided by smell, memory or some sixth sense like a homing pigeon, who knows for sure – my guess, for what it is worth, is that all three played a part, but for now the otters were in a territory which was familiar and safe. It was one of those summer nights when the stillness was all-pervading, amplifying the occasional sounds. The two used their ears as much as their eyes to alert them to the world around, knowing enough to tune out the normal. A rapid burst of frenzied biting, as a water vole separated a stalk, followed by the rustling reeds as it was dragged away to line the nearby burrow, went ignored, as did the occasional hoot from an owl. It was now a long while since the last glow of the sun had disappeared, and the wind had dropped to next to nothing. For a moment they paused as a single chime from

the squat flint-faced tower of our Norman church reverberated across the valley. The first hour of a new day was over,
but it was still night and I was alone, moving along the river
bank. Even though I had no torch I navigated with ease,
crossing bridges, climbing stiles and passing through the
woods, for it was one of those blue nights. I can't explain
how the cosmos creates such a thing, but that night was
typical of a time of year when it never really gets dark, even
though there is only a sliver of moon. The sky remained a
steel blue, a few static clouds breaking the patina, as did some
bright stars low on the horizon.

As I went along I couldn't pretend I saw everything in
crisp high definition; occasionally I stumbled on the uneven
ground or a tree root. The world around was slightly
hazed-out, colours hard to discern. It was all rather black and
white. Around me the bats were my aerial companions. In
the stillness they fluttered rather than flew; they reminded
me of butterflies gathering around a buddleia bush. I guessed
from the way they rose and fell that there must be some
localised thermals, as the air felt warm, but the damp, dew-
moistened grass felt clammy cold to the touch – my feet were
chilled and wet already. Bats have a fearsome reputation – all
that talk of vampires and Halloween depictions probably
doesn't help, but you have to get used to them, however low
or close they come, because, frankly, there are so many of
them, and if it is any consolation I've never had one hit or
touch me yet. Against the sky the little figures were inky
black, the fast-moving silhouettes with wings so bat-like that
you had to remind yourself that these were the real thing.
But the strangest thing was the lack of noise. No flapping.
No squeals. Nothing. Complete silence. However much they
wheeled, whisked and turned about you, it was like watching

the world from behind a triple-glazed plate-glass window; you know in reality the air is filled to bursting with the rapid-fire high-pitched squeaks that they emit for sonar navigation, but your ears tell you something different. I stopped for a moment to strain to hear anything, even the smallest indication, but I got back precisely nothing. I wondered for a moment if otters could hear the sound. If they did, Willow and Wisp didn't show it; in fact, in all the time I have ever watched otters I have never seen them glance up or even acknowledge the presence of the bats in any way. For them the world exists at eye level and below. Everything else is superfluous.

The big food items – fish, eels and so on – were still a struggle for Willow and Wisp. One fish in every three that is hunted is considered a good strike rate for a battle-hardened otter. The two were closer to double-figure probability unless they struck lucky, but they were not going to starve. At the height of summer the valley is thriving with life. Every creature that is ever likely to have offspring has had them by now. It is common to see a mother mallard with a convoy of a dozen ducklings in tow (this might already be the third brood since the spring), though in a week I can guarantee that the convoy will be depleted to two or three. Willow and Wisp are no innocents in this decimation.

Ducklings are easy prey. They can't fly, they can't hide very well and they certainly can't outswim an otter. Come dusk, the mother will gather them all together in a dense patch of reeds – they have long since abandoned the nest, and this new place has no strategic value beyond the hope that they won't be noticed. But the muddy margins of a river are most definitely the otter's domain. Whether Willow and Wisp actively seek out the ducks during their nightly patrols or simply stumble upon them, I don't know, but when they are found it is mayhem, as mallards have an extraordinary defence mechanism. The mother doesn't stay to defend the ducklings but rather creates a diversion, offering herself up as a sacrifice, heading off in one direction whilst the ducklings head in the other, dispersing as fast as they can. She'll quack and flap as if injured, hoping that the attacker would prefer one large meal to many small ones. It really is quite a commotion. If you didn't know it, you would think she had a broken wing, as she whacks her good wing down, thrashing up the water as she spins in ever-decreasing circles, for all the world a crippled bird.

But it never seems to work. I'll hear frantic cheeps from the panic-stricken ducklings from amongst the reeds that are pushed over as Willow and Wisp chase them down. Occasionally a duckling will pop out into open water in a bid for freedom, the little stubby wings propelling it forward, butterfly swimming-style, with remarkable speed for something so small. But it is mostly a short-lived escape as a muddied otter emerges, jaws at the ready, snatching it with little difficulty. In fact, the only respite the ducklings have is when the otters pause to swallow a brother or sister down whole. Head, beak, body and legs gone in a few choked-down bites. Eventually the commotion abates. Willow cruises along the reed line, occasionally stopping to investigate a perceived sound or movement, whilst Wisp creeps along the bank. They clearly know there are ducklings left, as all the while the mother, all pretence of injury now gone, keeps her distance, emitting a low metronomic quack every five seconds or so. The sound punctuates the dark. It is rather restful. Calming, even. A few downy little feathers drift off with the current. If there are ducklings still alive, they know better than to respond.

It is tempting to think the mallard is in a lament, the slow quacks a dirge for a family destroyed, but really I think it is a warning. A message. 'I'm here. Keep your head down. Be patient. Don't say a word or move a muscle.' Quack – pause – quack – pause – quack. On and on she goes. No little cheeps echo back in reply. Willow and Wisp start to get bored, sniffing the air. The mallard is starting to have a soporific effect on them as they pause to groom, Willow joining her sister on the bank. After a while they must conclude that whatever ducklings were left have either gone or that their efforts are better expended elsewhere, for with that loping half walk, half run they disappear into the night, a few coughs and eeks marking their progress until all that is left is silence. As they go, the mallard has stopped her noise, swimming level with the reeds, cocking her head to listen to the night. Satisfied, she emits one tiny quack, different to the ones before, and an even tinier cheep responds, drawing her into the heart of the reeds. She pushes her head beneath a matted clump, battered down in the mayhem, uncovering a bedraggled and terrified duckling which she steers behind her with her beak. They go in search of others. Another quack, another tiny cheep. The little group grows, zig-zagging amongst the reeds behind their mother until there is no more response. The dozen or so ducklings are now seven. It has been a bad night as the mother duck gathers them up and the much-depleted convoy heads off to find a safer billet. But safer is a relative word; the clock is ticking on that week where the countdown is on to two or three. They will be assailed from every angle; red kites in the air, pike from below, and no doubt Willow and Wisp will return for another foray. It is no fun being at the bottom of a food chain.

Whenever I saw Willow and Wisp it was the word 'companionable' that kept popping into my head, especially during those last few balmy summer nights when they struck out alone as Kuschta slowly lengthened the apron strings. For the most part, the pair took life at an easy pace, safe in the knowledge that their mother was never far away, food was plentiful and they were kings of this particular hill. They ambled rather than ran. They didn't compete, but rather deferred to one another. If one stopped to hunt a pool, the other would stop as well but would watch rather than join in, rolling in the grass or grooming to pass the time. If something was caught, there was no attempt to steal it or even to indulge in forcible sharing; the one would simply wait for the other to finish, maybe lick and sniff at a few scraps more out of curiosity than hunger, before they headed off in unison, trotting side by side or amiably swapping the lead.

But nothing is for ever. The summer would end. The winter would be hard. That much Kuschta instinctively knew. Her survival would be threatened by her own pups. It was time for them to be gone.

CHAPTER 14

THE LONG JOURNEY

Grayling and coots

Mion had never really gone away in the past year, that much Kuschta knew. Every so often she would come across his scent, or some indication that he had passed through. Occasionally she had heard him go by in the night but she'd kept hidden, with no desire to show herself or the pups. In the early days after their courtship and whilst she was pregnant, his comings and goings had been fairly common, but once the pups were born, less so. That was perhaps more a reflection on the constrained nature of her travels as she nurtured the newborns, or maybe Mion thought his presence unrequired. After all, he had three or four other Kuschtas to attend.

Actually, nobody has ever really got to the bottom of the relationship between the father and the family in the world of otters. Some observers have filmed the male keeping a nightly vigil outside the natal holt whilst the mother hunts. Others have seen some friendly interaction between both parents with the pups. Sometimes the mother has been seen to take great umbrage at the approach of the father, however innocent or well-intentioned the visit may be, launching a fierce attack to repel him in double-quick time. It must be said that the males do back off – there is no recorded case of a serious fight ensuing. However, at the far end of the spectrum, paternal infanticide demonstrates something far more sinister. As with so many things otter, there are many secrets they still keep from us.

With each passing month since the arrival of spring, being a mother had become easier for Kuschta. The winter had been very tough. Sustaining herself, plus Willow and Wisp, had been hard. They had scoured every corner of their territory, even encroached on that of others and exploited every bit of food that took them to the extremes of the things that otters really prefer to eat. No bit of carrion was left untouched. No living thing that might provide even the smallest protein hit was left uneaten. Otters might be apex predators but one of the secrets to their survival, not just as individuals but as a species, is a remarkable ability to eke out an existence in the toughest of times. The only avenue I haven't seen or heard of them going down, unlike, say, foxes, is rummaging through bins.

You certainly get used to living with otters. They become part of your life in a way I never imagined might happen to me. But this is not a two-way relationship. They have no interest in me, however fascinated I became by them. I am simply something to be ignored or avoided as they go about their lives. By the height of the summer they had me on a short leash. The commentariat talk about dog-whistle politics – this was dog-whistle wildlife and I was the canine. Every night I'd make a point of sleeping with the windows wide open. Actually I'd doze rather than sleep until that first eek or chirrup brought me instantly awake. Then I'd lie there listening for the next sound. Sometimes it never came. Maybe I had heard something else? But mostly it did. A splash. A surfacing cough. The churning water of the hunt. Another eek, perhaps, or the sound of running over the dry turf. On exceptionally still nights even a cacophony of crunching flesh percolated through the air.

The longer the summer went on, the bolder they became. They came earlier, stayed longer, ate more. They didn't make any pretence at concealment. No longer was the catch hauled off to the meadows but was eaten right here, on the edge of the lake. Soon there were three or four distinct slides with the patch of grass dead, turned brown by the spraint and urine of regular use. Oftentimes all three would be tucking into their respective fish simultaneously. It wasn't the best of times to be a trout, as Willow, Wisp and Kuschta camped out in the meadows during the day, ready to return just as soon as the fall of dusk allowed. And then suddenly nothing. No raiders in the darkness. No daytime evidence of night-time kills. Even our home-made holt had that feeling of desertion, the once-trampled entrance path now sprouting covering growth. Something had changed.

The pups may have been growing and life was on an upward trajectory for them, but for Kuschta it was going full circle. Until now, any signs of Mion had raised her hackles, creating suspicion and aggression in equal measure. It was family first, Mion nowhere. However, that faded as his scent began to tweak gently at her libido. She sought out his spraints, taking a sort of comfort from his presence and sensing a change within her. It was time Kuschta tried, albeit by gentle degrees, to rid herself of Willow and Wisp. Sometimes she deliberately separated herself from the pair so that they spent the day and occasionally the night apart as she ranged to the furthest extremities of their territory. But that was really the problem. All three were confined by the invisible boundaries of their homeland. The pups would never stray from the familiar and Kuschta couldn't impinge on alien territory for any length of time without considerable risk. So, one way or another, they were never apart for long. The reunions (for Kuschta tried this tactic more than once . . .) were a joy for Willow and Wisp, who tumbled over and around their mother with delight, to which she in turn acquiesced rather than enjoyed.

You'd have thought that at some point Kuschta would have lost her patience, pushing her offspring to the edge by definitively withdrawing her love and protection in the way that she had done to cause the death of that four-month-old pup. I guess it would have been pointless; after all, both her daughters were by now almost self-sufficient. However, nor did she chase them away like she did Mion, but then again maybe it is in the male disposition to disperse, in order to mix up the gene pool – all he needed was the motivation. Frankly, it seemed the mother-daughter bond was too ingrained to be broken by anything she had tried before. It was time to do to her own daughters what had been done to her.

The long journey started out on a night like any other of that summer – Kuschta intended it to be that way. As the sun faded from the sky the otters stretched and itched after a long, idle day at their various couches and resting places. Padding along well-worn paths, they came together at the Willow Island, a low whickering announcing their arrival as they gathered together, sniffing and nosing at each other in familial greeting. For a while they groomed, in no great hurry to move. Willow splashed into the brook, pawed about in the shallows and was rewarded with a small eel for her trouble.

Wisp and Kuschta showed no signs of being similarly inclined to hunt. Nor did they show any apparent envy at her success, watching with mild curiosity as she ate. And then, as is the way with otters, they went from idle to action in a moment, Kuschta jumping up and heading off downstream with no prior warning. Wisp immediately fell in behind, whilst Willow had a moment of indecision, grasping the remains of the eel between her forepaws. In a sort of double take, she looked from the eel to the departing figures and then back to the eel. Whatever calculations go on in the head of an otter at moments like this, I have no idea, but for Willow the thought of being left behind was clearly worse than forgoing a few last mouthfuls. She dropped the eel without further hesitation, sprinting to catch up.

For the next few miles the trio charted familiar territory. The Badlands gave way to the open meadows. They passed by where once there had been a muddy slide, Kuschta's path in and out of Beech Wood. Nobody showed any inclination to pause, still less revisit the holt of their birth. Had they taken the time to mount the incline and push their way through the undergrowth, they'd have found the foxglove hollow temporarily occupied by a family of foxes, who delighted in the home as much as they had done. But on they went, less inhibited as they gained the cover from the thread of woodland along the Wallop Brook as it flowed on beside the ripening wheat fields, dark trees marking out the skyline along the ridgeway high above.

Soon the landmarks started to run out. At the fork in the river Willow and Wisp paused on the junction as Kuschta continued on downstream. For a while they waited, confident that she would return to lead them, if not back home, at least up the other fork in a direction similar to home. But Kuschta kept on her way. She knew the pups would follow in the end so she charted her progress with the occasional eek that became ever more distant until, fearful of the unknown, the pups scurried in her wake. By now the brook, as it approached the junction with the main river, was gathering a little more size and pace so they swam rather than walked. The three slid over the concrete sill at the water gauging station, and as they passed under the single arch of the red-brick bridge on Horsebridge Lane they had now crossed into alien territory.

Carried by the current, they let the water do the work; swimming is so often effortless for them. The head up, the body largely submerged and that flexing tail so well described as the rudder, they kept together in close formation until they were swept into the River Test. Liberated by the much larger river, Willow and Wisp dived and frolicked in the competing flows at the confluence, going deep where the back eddies had scoured holes. They disturbed fish much larger than anything they had ever encountered in a river before, their whiskers humming with vibrations, encouraging them to plumb the depths dive after dive, but the fish, wise to the ways of otters, edged away to safety. Kuschta left them to it, cruising the margins, and within a very short time surfaced with a fish, dragging it up onto a mid-stream island. A few sniffs told her this was no place to linger long; it belonged to others. So when she was joined by her unsuccessful daughters they shared the trout in rapid-quick fashion. It was time to move on.

Kuschta was determined to cover a great distance. Over the next two nights she wanted to find a new place far, far away, so unfamiliar and different that Willow and Wisp would never find their way back to The Badlands. While it was still dark, she didn't let them rest. They didn't pause to hunt. She urged them on when they faltered. They were moving upstream, away from the sea, in the opposite direction to Lutran. It was an easy landscape to traverse, flatter and more open than the Wallop Valley. Long ago the banks were largely denuded of trees by Victorian anglers who preferred it that way for their sport, with open meadows stretching way into the distance. Today it is not much changed, with mile upon mile of manicured bank, the grass trimmed for the convenience of fly fishermen and with few obstructions for people, let alone otters. They moved quickly along this riparian super-highway, choosing land over water for more speed at less effort. Occasionally a building or some other obstacle stood in their path, forcing them into the river, but for the most part they kept dry.

By 4am, with the first chink of pre-dawn turning the white clouds pink, the pups were fading for lack of rest and food as Kuschta guided them off the main river up a tributary almost as large and fast as the main river itself. Ahead stood Eight Hatches, where the water pounded out from between nine concrete uprights, creating an octet of separate waterfalls feeding a giant, churning pool. From the river, Eight Hatches was impassable for the otters, but that didn't matter – Kuschta had no intention of going any further that night. Naturally the pool was tempting, but she had her eyes on the gravel shallows immediately below, a vast mattress of loose stones spat out by the erosion from the torrents and home to shoals of grayling.

A grayling is something of an oddity, a fish that doesn't quite fit into the natural order of a chalkstream even though it has been here since the Ice Age. It is a member of the salmon family but doesn't migrate. It is related to the brown trout but spawns in the spring rather than the winter. It prefers the company of its own brethren rather than the solitary life.

And it has a look and a smell unlike any other river fish, a huge red-edged dorsal fin atop a hard-scaled, steel-grey body with a forked tail and a snout rather than a mouth. The smell? Well, it is distinct. River keepers of old used to claim they could get a whiff of grayling on a cold, frosty morning. I've never sniffed the dawn and thought 'Ah, grayling' myself, but, once handled, your hands will have a peculiar odour. Not unpleasant, but certainly different. It is not fishy. You'll search your olfactory memory bank for a clue. If you know your herbs it will hit you: thyme. And you'd be right. The Latin name for grayling is *Thymallus thymallus*, named after the herb plant by Swedish zoologist Carl Linnaeus in 1758.

Otters have a particular fascination for grayling, seeking them out in preference to everything bar eels. Maybe the smell aids the hunt. Perhaps the firm flesh is particularly tasty. Whatever the exact reasons – I am sure they are myriad – Kuschta knew that, come the dawn, the grayling would gather on the shallows to feed. Every generation would be there; tiny fingerlings come out from the slack water, big specimens from the deep of the pool and everything in between competing for the freshwater shrimps and larvae that tumble through the hatches. On the shallows the three otters resemble bears on a salmon river as they half swim, half wade after the grayling. Heads plunge beneath the water, shortly to reappear with a flapping fish snared in the mouth. Occasionally there is a chase as one of the otters runs after a panic-stricken wriggling fish that has unwisely chosen an escape route that takes it into ever shallower water until stranded, where sharp claws and bared teeth deliver the denouement.

The smaller captures are eaten *in situ*. Larger fish are taken to a cattle drink where the hooves have created a gently sloping, crescent-shaped beach. Back and forth they went, all getting their share. Kuschta had chosen well. Grayling don't spook like trout, the shoals remaining intact despite the commotion and the gradual depletion of their numbers. In fact it was sunrise that put an end to the feast, as the brightness of a new day sent the grayling back whence they had come. The grayling would return at dusk to repeat the whole performance.

For a while Willow and Wisp mooched about in the shallows, missing the action, filling in the time by pawing over stones and surveying the surface for signs of non-existent fish. Kuschta knew better than to bother, grooming on the beach until she gathered them back with a low whicker. Up the slope and across the meadow they headed. It was really too bright and too late for them to be on the move, so they stuck to the cattle paths, shielded from the outside world by the tall meadowsweet with its creamy lamb-tail flowers and the purple-headed thistles. The grass was wet with the morning dew. All about them the spider webs glistening with necklace beads of water, the night-time captures, the struggle for life long over, now inert and cocooned in pale white filament ready for consumption some-time soon. As the three lolloped along, they left a slight aroma of menthol in the air as they crushed wild mint beneath their feet, but for the most part the only evidence of them passing this way was their dark trails in the gleaming silver of the dew. Ahead lay a scruffy clump of alders around a wet sump of water, largely shielded by nettles and brambles. A few cattle stood around the edge, hogging the shade, noisily ruminating as tails flicked away the first fly swarms of the day. The otters didn't pause for them as they headed into the thicket, but showed enough respect by keeping out of kicking range. Once inside, they halted amongst the gloom, the earthy dampness comfort-ingly familiar. Kuschta pushed into the dry base of the brambles, whilst Willow and Wisp curled up together amongst the tree roots. Sleep came quickly.

A few bright stars shone out from the night-blue sky as Kuschta led her daughters out for the last time. The cattle had moved away, somewhere out there in the meadow grazing in the cool of the night, the noise of rasping breathing punctuated by a tearing sound as they wrapped their tongues around tuft after tuft of grass and clover. Kuschta was tempted to head back to Eight Hatches for an easy meal, but it was enough for now to know that Willow and Wisp had been taught something that would stand them in good stead. No doubt in the months, and maybe years, to come they would find their way back to the grayling shallows themselves, but for now their future lay further up river.

The Plantation was somewhere Kuschta knew well. Once it had been her home: a huge reed bed running to many hundreds of acres through which the river threaded, flanked by a complex labyrinth of side streams and ditches that made it almost impossible to traverse for all but the most purposeful of humans. For centuries this had had a rationale, the place where reed was harvested to roof the cottages that populate the villages that dot the valley. Today the roofs look quaint, with little eyebrow windows popping out from the sloping pitch and complex weaves along the roofline that are the signature of each individual thatcher. But really it was a practical thing. The reed grew in profusion, it was on the doorstep and did the job. Why not use it? Today, less so. Straw is a more common material, and reed, when used, finds its way from Hungary and eastern Europe.

The abandoned reed beds are not entirely left to nature. Tall cricket-bat willows grow straight and tall, planted in geometric lines. The trunks are bare of all side branches; just the top, fifteen or twenty feet up, sprouts green branches. Of course they don't happen this way by accident. Planted as branchless, leafless, single, straight spars, they are more akin to a dingy mast than a tree. The 'sets', as they are called at this stage, don't require a hole as such for planting; each is no thicker than your wrist so a crowbar pushed three feet or so into the boggy ground is a sufficient opening to establish these water-loving *Salix*. It might be tempting to say that is it; just sit back and wait fifteen or twenty years until the tree can be felled, cut into four-foot planks ready to shape into those famous willow cricket bats. But it doesn't quite work like that. The first few years are critical as a neglected tree will have no commercial value. Side branches and shoots create knots in the pale white wood, making it useless for any other purpose than charcoal burning. So as the side shoots appear they are rubbed off with a gloved hand in the early years then trimmed in later years so the tree is forced to grow higher and higher, creating that clean, straight trunk. Eventually, when you can stand beside the tree and encircle it with your arms so that your fingers touch on the other side, you'll know it is time to call in the saw.

For the occasional willow trimmer the Plantation was as much an abandoned landscape as The Badlands, and one over which otters will always hold sway. No person or creature really bothers them, the reeds too dense and the ground too wet. This is the home of water voles and wildfowl, frogs and toads, snails and water-loving bugs, tiny fish and freshwater crustaceans. And as a place to abandon Willow and Wisp? Well, nowhere would ever be perfect, as all the best territories are claimed by someone. That much Kuschta knew, but here was a place where they had a chance. In many respects they were luckier than Lutran. The world is a hostile place for a juvenile male; nobody likes him. Nobody wants him. He is either a threat or a nuisance, another mouth competing for scarce food. But young females excite less antipathy. Dominant males regard them as you might expect – as future breeding stock. Other females tend to be pragmatic, allowing for over-lapping territories just so long as there is enough space and food to go around. Leaving Willow and Wisp would be Kuschta's final act as their mother. It would not be gradual. It would not be anticipated. One moment she would be there, the next gone. And in that final day and night she did nothing to hint at what was soon to come.

As they emerged from the river, paddling through the muddy margin towards the tall curtain of reed beyond, a group of alarmed coots followed the trio, squawking in their wake, the sound loud in the quiet of the night. Just inside the reeds there was a nest sitting well above the water line, a circle of raggedly woven reeds atop a pile of longer reeds arranged in a random crisscross fashion as if thrown down in preparation for a game of pickup sticks. At first glance coots don't appear to be great ones for natal engineering; even a nest in regular use has an air of disintegration, seemingly about to collapse at any moment, but in fact they are surpris-ingly durable, surviving most of what nature throws at them long after the nesting season is over. For now, this one was most definitely in use, on maybe the third, possibly the fourth, clutch of eggs since the spring. Why so many broods? Well, coots are even more attritional with their young than otters; the average laying of nine eggs will be whittled down to three within a week of hatching. Even in the best of times the parents can't harvest enough food to feed all the hungry mouths, as they rely on hard-to-gather shrimps and insects to sustain the early stages of life. So, unable to satisfy the demands of their mewling offspring the parents start to attack the chicks for no other reason than that they have had the temerity to beg to avoid starvation. At three or four days old it becomes clear that some, actually most, of the chicks must perish. The adults become brutally pragmatic, concentrating their ire on the weakest, who simply give up begging, preferring to die of starvation rather than accept regular beatings. For such an innocuous-looking bird, the coot really does have a pitiless way of selecting the fittest to survive.

Back on the nest it was near impossible to see the black-feathered body of the mother coot but for the white top of her head that caught the reflection of the moon as she cocked it this way and that with increasing agitation as the otters neared. With little defence beyond remaining put, she began to spit staccato squawks in the direction of the advance whilst the flock behind became increasingly shrill. You have to give it to the coots; they tried their best. But a few beaks, a bit of noise and some flapping wings is no match for three otters. The mother let out one final scream as Willow nosed at the base of the nest before scrabbling away at speed, those spindly legs scooting her across the surface of the soft mud into the night, followed by the others who clucked and fussed in abject defeat as they disappeared into the distance.

The nine or ten eggs were just a day or two away from hatching, but the otters neither thought nor cared as they clambered around the nest, gradually pressing it into the ooze by sheer virtue of their weight. Each taking their share, they bit into the shells, licked away the albumen- and blood-specked yolk that coated the unhatched chicks before crunching into the twitching bodies as feathers, legs, beaks and flesh disappeared in a few swallows. It was all over in a minute or two.

A little later the coots returned. They fussed and clucked

around the scene of devastation as if in discussion about what had occurred, pecking at the shards of eggshell and chick remains. Well, no nutrition must ever go to waste. There is no sentiment in the animal kingdom. After a while, whatever had gone before seemed to have been forgotten (or maybe coots are essentially pragmatic), the conversation ending as they busied themselves, gathering hollow straws of reed, snapping and dragging them into position for a new nest. Nothing would be gained by not trying again.

For Kuschta, in that final night with Willow and Wisp, there was little more she could teach them. At close to fifteen months old they had really gone beyond the normal time together. The best she could do for them now was show them the extent of the Plantation. The boundaries. The resting places. The otter highways. The holts. The little-used extremities where they might be safe. The year ahead would be full of dangers. Death would be ever-present as man, cars, fights and starvation took their toll. Bridge this time from adolescence to adulthood and there was a future, but in between the risks were huge. As nomads trying to settle in an unfamiliar land, every day would be a challenge.

If they made it through, Willow and Wisp would become true apex predators. The threat rather than the threatened. In the parlance of otter hierarchy both would become 'landowners', establishing a territory to call their own. A home range where they would welcome, for a short while, a mate, and then raise a litter all of their own. A new generation depended on their success.

But for now, as they settled down for a long summer day of rest, that was a future unknown.

EPILOGUE

At dusk Kuschta left. There was no final parting. No last moment together. She simply vacated her couch, pushed through the reeds, slipped into the river and was gone. As she swam downriver in the ever-darkening gloom, did she recall her first night alone? The fright. The confusion. The enormity of unasked solitude. Did some maternal urge scream at her from within to turn and return?

It would be human to hope so, that Kuschta spins around and goes back to her daughters. She knows that a few more days will make little difference to the final outcome of their lives, but she takes the chance to grab some joy from what is mostly sad. The Plantation becomes the epicentre of their world. A place they can truly call their own. They pull eels from the mire. Creep up on love-struck frogs. Sniff out long-abandoned holts, snuggling together for contented rest. They raid nests. Snag fat trout. They want for nothing and fear nobody. For all the pain of the upcoming separation, they frolic and hunt as if tomorrow will never come.

But of course none of this happens. Kuschta doesn't turn around; she is impelled by something far greater. Willow and Wisp are on their own, for good or ill. For the first few days they would cling together, united in the confusion of being abandoned by their mother. But quickly the solitary otter gene would raise its head as the two separate to carve out lives and territories of their own. In less than a week the most cohesive family unit in the animal kingdom has been broken asunder as one becomes three. It is a sad end to a wonderful thing.

For two nights an unencumbered Kuschta travelled fast, driven by that deep stirring inside her. Constantly on the move under the cover of dark, with little desire to stop bar the occasional need to hunt, she passed unseen through alien territories. At Eight Hatches she paused to trawl the deep pool but didn't hang about for the grayling. In her impatience to stretch the limits of the night she swam on well past dawn, using the early morning mist for cover until the sun burnt it off. All day she was restless, fidgeting on some unfamiliar couch. The arrival of night was a welcome relief.

On that second night she didn't stop. She didn't need to. She was going home. As the miles fell away, so the familiar landmarks appeared, every twist and turn of the Wallop Brook telling the story of her life. At The Badlands she finally came to a halt, exhausted but content. Sniffing the wind, a long-forgotten scent came her way. Mion could not be far away, that much she knew. Kuschta settled down to wait. Her time had come again.

BIBLIOGRAPHY

Adams, Ena et al. *Deer, Hare & Otter Hunting* (Lonsdale Library: Volume XXII) (London, Seeley, Service & Co., 1936)

Allen, Daniel. *Otter* (London, Reaktion Books, 2010)

Chanin, Paul. *Otters* (Stansted, Whittet Books, 2013)

Chester, Nicola. *Otters* (London, Bloomsbury, 2014)

Conroy, J. W. H. et al. *A Guide to the Identification of Prey Remains in Otter Spraints* (London, The Mammal Society, 2005)

Harris, C. J., *Otters: A Study of the Recent Lutrinae* (London, Weidenfeld and Nicolson, 1968)

Kruuk, Hans, *Wild Otters: Predation and Populations* (Oxford, Oxford University Press, 1996)

Laidler, Liz. *Otters in Britain* (Newton Abbot, David & Charles, 1982)

Lomax, James. *Otter Hunting Diary: 1829 to 1871* (Blackburn, Thomas Briggs, 1910)

Stephens, Marie N. *The Natural History of the Otter: The Otter Report* (Potters Bar, The Universities Federation for Animal Welfare, 1957)

Williams, James. *The Otter* (Ludlow, Merlin Unwin, 2010)

ACKNOWLEDGEMENTS

The acknowledgments are usually reserved for those both professional and personal who helped bring the particular book into being. But on this occasion, I need to thank a group of people I have never met or known. Namely those who tirelessly campaigned to save the otter from extinction.

As I write this approaching 6pm in the dark of a cold January night where the frost is already gathering on the ground, I can hear the eek of approaching otters. It is a noise that has become the soundtrack of my life but it could so easily not have been.

For without those campaigners, who fought the forces of government, business, ignorance and indifference with little thanks or support over five decades, the English otter would have been consigned to a natural history footnote.

But against huge odds they succeeded so we may all collectively reap the reward in the steady revival of this truly remarkable animal. We owe those people a huge debt. My thanks are inadequate but they are sincere. Thank you – this book would never have existed without you.

Nether Wallop Mill, January 2017

INDEX

Acte for the Preservation of Grayne (1566) 49, 49n
agriculture 48, 52−4, 57, 122−3, 143−4, 150
apex predator 19, 135−42, 163, 228, 248, 260
Armand-Delille, Dr Paul-Félix 150
autumn 9, 29, 62, 127, 133, 140, 153−70, 177, 178, 180, 211−12, 217, 226

Badlands, The (area downstream of Nether Wallop Mill) 44, 46, 57−60, 63, 68, 76, 79, 92, 93, 94, 95, 103, 106, 111, 116, 124, 141, 145−6, 182, 188−9, 216−17, 240, 251, 252, 257, 262
bats 21, 33, 111, 134, 242−3
Beech Wood 77, 80, 83, 84, 86, 87, 90, 93, 95, 146, 229, 251
blue water-speedwell (Veronica anagallis-aquatica) 159, 160
bullhead (Cottus gobio) 129−32, 135, 147, 204
bumble bee 91, 177−8

Churchill, Winston 187, 188
common brown rat (Rattus norvegicus) 219−20

coots 131, 139, 142, 258−60
couches (concealed resting places) 6, 7, 12, 29, 41, 42−3, 59, 70, 134, 145, 165−6, 191, 192, 197, 200, 215, 238, 240, 250, 261, 262
crayfish 23, 57, 98, 107−9, 125−8, 136, 145, 173, 174, 179, 197, 199, 203, 214−15, 238

ducklings 105, 132, 135−6, 243−5

eels 101−4, 106, 107, 124−5, 145, 179−81, 203, 243, 250, 251, 254, 261
Eight Hatches 253, 256, 262
estuary otter 231
Eurasian or European otter (Lutra lutra) 10−11, 19, 68
European water vole (Arvicola amphibius) 140−2
extinction, possibility of otter 47, 48, 51, 54−5, 56, 236

Follo, Roger 49
frogs 21, 105, 106, 132, 147, 155, 179, 181−6, 257, 261
fyke nets 173

game fish 204
Grahame, Kenneth: *The Wind in the Willows* v
grayling (*Thymallus thymallus*) 131, 204, 253–5, 256, 262

hedgehog tick (*Ixodes hexagonus*) 168–9
Henry II, King of England 48
Henry VI, King of England 49
holts 5–7, 10, 66, 70, 71, 74, 75, 77–9, 80, 81, 82, 83, 86, 87, 88, 92–4, 95, 104, 124, 134, 154, 165, 171, 191–8, 200, 203, 213, 215, 221, 229, 248, 249, 260, 261
  author builds 192–8, 213
  birthing/natal holt 75, 77–9, 83–4, 87, 92–4, 200, 248, 260, 261

Ireland 47, 109, 126
Itchen, River 153, 236–7

Keats, John: 'To Autumn' 153, 155
Kendall's coefficient of concordance 106–7
King's Otterer 48–9
Kuschta (female otter):
  abandoned by mother 5–10, 11–13, 18
  abandons cub to die 109, 113–16, 117, 239–40, 250
  abandons cubs to live independently 211–14, 239–40, 246, 248, 249–60, 261, 262
  age/length of life 176
  birth of cubs and 80, 81–2

dogs and 58–9
early weeks with cubs 85–7
feeding of cubs 83–7, 88–9, 90, 92, 101, 109, 112, 113
grooming 100–1, 255
hunt, teaches cubs to 98, 99–100, 112–14, 125–32, 144, 146–9, 180–6
hunting 11, 13–17, 21, 22, 23, 32, 34–6, 70, 83–5, 86–7, 89, 92, 98–100, 101–4, 111–13, 118, 123, 125–32, 135, 149, 154, 180–6, 199–200, 202–4, 209–10, 211, 214–15, 217–21, 227–8, 233, 236, 243–6, 254–5, 259–60
leaves cubs unprotected to hunt 83–5, 86–7
mating 63, 66, 68–75, 247, 249–50, 262
Mion and *see* Mion
move from birthing/natal holt 87, 92–4, 95–7, 101
name 12–13
natal holt 77–9, 83–4, 87
newly born cubs and 81–3
search for new home after abandonment by mother 17–23, 29–36, 45–6, 57–63
solitary life 19, 62–3
swim, teaches cubs how to 97–8
territory, size of 65
trout lake, Nether Wallop Mill, visits to 60–2, 79, 80, 83–5, 86–7, 88, 89–90, 110–12, 144, 209–11, 233–4

weight 144
weaning of cubs 88

lobster pots 109, 173−4
Lutran (son of Kuschta):
appearance 118, 154, 180
couch 134
early weeks of life 93, 118,
119, 123
ejection from family group
170, 209, 211−14, 257−8
fights Old Dog 221−3
future of 180, 231
hunting 123, 125, 128, 131−2,
141, 148−9, 181, 184,
214−15, 216−19, 220−1,
226−9
search for new home after
ejection from family
group 215−31, 239, 257−8
unruly/disruptive, becomes
154
weight and food consump-
tion 144

Magna Carta 50
Meon, River 205
Mion (male otter):
age 66, 176
arrival of Kuschta in Wallop
Valley and 65−6
challenges to authority/
fights for dominance
66−8, 164−5, 169−70
cubs and 75, 78, 164, 190,
247−8
females, reaction to arrival
in territory of new 66, 170
hunting 161−2

grooming 166−9
mating duties with various
females 164
mating with Kuschta, first
time 68−75
mating with Kuschta, second
time 211, 247, 249−50, 262
Nether Wallop Mill and
160−1
range of travels 161, 163, 168
roles 63, 65−7, 164−5, 247−8
summer and 163, 164−70
sunbathing/resting 165−6
territory and family groups
162−4, 169−70
ticks and fleas and 168−9
mink 41, 54−5, 82, 83, 136−9,
141−2, 172, 238

Native American tribes 12−13
Nether Wallop, Hampshire 44,
107
Nether Wallop Mill 110−11,
160−1, 169, 178, 191, 192,
213, 233
author buys 37−9
author first otter sightings
at 40−5 *see also* trout lake
noise of water at 109−10
river flow maintenance and
210−11
The Badlands and 46
trout lake 44−5, 60−2, 79,
80, 83−5, 86−7, 88,
89−90, 110−13, 115, 116,
144, 192, 212, 209−11, 213,
233−4, 249
North American otter 192

Old Dog (male otter) 222–3
organochlorines 52–6, 57, 151
organophosphates 56, 57, 151
otter:
  apex predator 19, 135–42,
    163, 228, 248, 260
  author's first sightings of
    37–42, 236–8
  breeding season 75, 164,
    235
  British folklore, history and
    culture, lack of presence
    in 46–7
  calls 68, 69, 72, 111, 113, 134,
    212, 236
  carrion, eating of 90, 105,
    248
  couches (concealed resting
    places) 6, 7, 12, 29, 41,
    42–3, 59, 70, 134, 145,
    165–6, 191, 192, 197, 200,
    215, 238, 240, 250, 261, 262
  cough 2, 16, 34, 42, 73, 85,
    110, 161, 225, 245, 249
  cubs
    abandoned by mother to
      die 109, 113–16, 117,
      239–40, 250
    abandoned by mother to
      live independently
      9–10, 211–14, 239, 240,
      246, 248, 249–60, 261,
      262
    birth of 81–2
    birthing/natal holt 75,
      77–9, 83–4, 87, 92–3,
      200, 248, 260, 261
    early weeks of life 81–8
    hunt, learn how to 9, 10,
      13–17, 98, 99–100,
      112–14, 125–32, 144,
      146–9, 180–6
    left unprotected whilst
      mother hunts 22, 83–5,
      86–7
    move from natal holt 87,
      92–4, 95
    swim, learn how to 97–
      8
    time spent with parent
      compared to other
      animals in same area
      (sympatric species)
      98–9
    water survival 93–4
    weaning 88
daily food intake 101, 144–5,
  190
dangerous life of 171–86
death rates 171–6
diet 12, 13–17, 23, 32–6,
  44–5, 53–4, 57, 60–2,
  79–80, 83–90, 98,
  99–100, 101–9, 110,
  124–32, 141, 142–9,
  161–2, 163, 167–8, 173,
  174, 178, 179–86, 192–3,
  197, 199, 203–8, 209, 211,
  214–22, 234, 238, 243–5,
  249, 250, 251, 252, 253–5,
  256, 261, 262 *see also*
  under individual type of
  prey
digestion system 200
diseases 175
distances covered 10, 18, 24,
  57, 161, 163, 168, 176, 200,
  252–3

dogs and 13, 57–8, 176
eating of prey 17, 36, 45, 62,
　103–4, 115, 125–8, 200,
　209–10, 228, 249
evolution of 25, 31, 66, 115,
　132, 135, 136, 159, 172, 208,
　221, 239–40, 259
extinction, possibility of 47,
　48, 51, 54–5, 56, 236
eyesight 105, 200
fat reserves 176, 190
fighting 20, 23–4, 66, 67–8,
　169–70, 175–6, 222–3
fleas and 150, 167–8, 169
food stocks, conservation of
　22–3, 25, 172, 199
fur 17, 27–9, 98, 100, 103, 116,
　118, 137, 138, 166–7, 168,
　172, 175, 176, 190, 198, 212,
　231, 235
gestation 75–80
grooming 17, 28–9, 33, 42,
　70, 85, 86, 100–1, 104, 113,
　128, 134, 166–7, 184, 198,
　199, 202, 212, 225, 228, 245,
　246, 250, 255
hearing 30, 200–2, 241–2
hedgehog tick (*Ixodes
　hexagonus*) and 168–9
hierarchies 24–5
　holts (tunnels) 5–7, 10,
　66, 70, 71, 87, 92–3,
　191–2, 203, 215, 229
　author builds 192–8, 213
　birthing/natal holt 75,
　77–9, 83–4, 87, 92–3,
　200, 248, 260, 261
　move from birthing holt
　87, 92–4

home, search for new/itin-
　erant months 10, 17–23,
　215–31, 239, 257–8
humans and
　eating of otter 47–8
　fyke nets/crayfish traps/
　lobster pots, otter and
　173–4
　habitat destruction 54,
　55–6
　history of interaction
　with otter 47–57
　hunting of otter 6,
　19–20, 47–51, 54, 68,
　93–4, 173, 241
　organochlorides and
　52–6, 57, 151
　organophosphates and 56,
　57, 151
　road deaths of otter and
　22, 123, 171–3, 241
　synthetic pyrethroids and
　56
hunting
　cubs learn 9, 10, 13–17, 88,
　98, 99–100, 112–14,
　118, 125–32, 144, 146–9,
　180–6, 202–4, 211,
　214–15, 216, 234, 243
　cubs left unprotected
　while mother hunts 22,
　83–5, 86–7
　Kuschta 11, 13–17, 22, 23,
　32, 34–6, 70, 83–5,
　86–7, 89, 92, 98–100,
　101–4, 111–13, 118, 123,
　125–32, 135, 149, 154,
　180–6, 199–200,
　202–4, 209–10, 211,

214–15, 217–21, 227–8,
233, 236, 243–6, 254–5,
259–60
mating and 70
mother hunts in daytime
during cubs early
weeks of life 22, 83–5,
92
solitary activity 70
stealth and 13, 32, 147,
180–1
see also under individual
type of prey
hunting of, human 6, 19–20,
48–51, 54, 68, 93–4, 173,
241
ice, crossing 240–1
infant mortality 171–2
infanticide 22–3, 78, 171–2,
248
land, happiness/movement
on 18, 31, 147, 198–9
length of life 171–6, 180
linear habits and outlook 10,
44, 169
mating 9, 19, 21, 23, 63, 66,
68–75, 160, 164, 249, 262
metabolic rate 165, 190
movement from land to
water 31–2
mustelid family and 24, 55,
99, 138
navigation 240–1
newly born 81–3
night, typical winter 191,
198–208
nocturnal nature 11–12, 19,
47, 61–2, 92, 111, 202
pesticides and 52–6, 57, 151

population numbers 50–7,
61, 115, 151, 172, 175, 238
reproductive cycle 9–10
road deaths 22, 123, 171–3,
241
runways, historic 241
seasonal adaptation, lack of
190
secretive nature of 47, 57
size of 46–7, 55, 67, 73, 82,
100, 137, 154, 237, 239
sleep 5, 6, 7, 9, 12, 18, 28, 36,
82, 83, 85, 92, 100, 165, 180,
191, 195, 196, 197, 198, 200,
203, 215, 220, 221, 228, 231,
256
smell, sense of 21, 68, 90,
102, 105, 174, 184, 186,
203, 217, 220, 229, 240,
241, 254
solitary nature 18–19, 25, 70,
62–3, 234–5, 253, 262
species 10–11
sprainting/marking territory
20–1, 22, 23, 25, 41, 42–3,
61, 79, 169, 192, 197, 213,
215, 229–30, 239, 240, 249
starvation 10, 51–2, 114–15,
123, 176, 179, 260
sunbathing/resting 165–6
swimming 3, 29, 32, 78, 93–4,
110, 227, 230, 237, 240, 244,
245, 252
territory
guard as food source 101,
176
humans invade 46, 55
juveniles travel great
distances to find new

5–12, 18, 57, 215–31, 239, 257–8
male range of 65, 163–4
mating and 70, 71, 75
mother abandonment of juveniles in unfamiliar 5–10, 11–12, 18, 211, 246, 249–60, 261, 262
natal home and 78, 79, 101
newcomers entry into 66–7, 169–70, 175–6, 222–3
resting places in, number of 197
marking (sprainting) 20–1, 22, 23, 25, 41, 42–3, 61, 63, 79, 169, 192, 197, 213, 215, 229–30, 239, 240, 249
time limit in water 29
tracking 29, 238–9
whiskers 13, 14, 15, 32–3, 34, 67, 69, 82, 84, 88, 105, 184, 198, 202, 203, 216, 237, 252
otter-hound pack 49
Otterer's Fee 49
Over Wallop 44, 161

pike (*Esox lucius*) 76, 104, 120, 121, 135–6, 137, 201, 204, 245
Plantation, The 256–60, 261–2
pollarding trees 124, 194–6

rabbits 33, 60, 68, 77, 80, 91–2, 95, 97, 98, 110, 122, 190, 223, 238
European rabbit (*Oryctolagus cuniculus*) 149–50
fleas and 167–8

myxomatosis and 150–1, 167
otter hunting of 105, 132, 142–9, 167–8, 178, 179–80
population numbers 149–50, 151

Sadlers Mill 225, 229–30
salmon 56, 104, 105–6, 130, 201–8, 225, 226–9, 253, 254
Scotland/Scottish Islands 11, 47, 78, 84, 115, 174
Scruffy (buzzard) 178–9, 193
sea otter (*Enhydra lutris*) 10, 11, 28
sloes 58, 189–90
solstice, winter 61, 187, 188
Southampton Water 231
sprainting 20–1, 22, 23, 24, 25, 41, 42–3, 61, 63, 79, 169, 192, 197, 199, 213, 215, 229–30, 239, 240, 249
spring 9, 36, 43–4, 45, 58, 62, 65–80, 85–6, 96, 107, 121, 143–4, 154, 155, 157, 177, 179–80, 182, 187–8, 189, 204, 208, 233, 243, 248, 253, 258
stoat 83, 87, 92, 110, 111, 143, 190, 238
Stonehenge 121, 187
summer 5, 9, 12, 29, 41–2, 43–4, 45, 60, 61, 75, 90–1, 99, 102–3, 110–11, 115–16, 117–32, 134, 136, 137, 144, 153, 155, 156, 157, 159, 163, 164–9, 189, 191, 192–3, 195, 197, 199–200, 211–12, 233, 236–8, 241–6, 247–60
synthetic pyrethroids 56

Test, River 119, 163, 175, 216, 224, 226, 252
trout:
 breeding season 130, 203–8
 farming 161–3, 217–24
 Nether Wallop Mill trout
  lake 44–5, 60–2, 79, 80,
  83–90, 110–13, 115, 116,
  144, 192, 212, 209–11, 213,
  233–4, 249
 otter hunting/eating of 12,
  13–17, 32–6, 44–5, 60–2,
  79–80, 83–90, 99–100,
  104, 105–6, 110, 145,
  161–2, 163, 181, 192–3,
  208, 209, 211, 217–22, 234,
  249, 252, 261

Valet of our Otter-Hounds 49

Wallop Brook 37–8, 43, 44, 45,
 51, 75–6, 77, 96, 98–9, 109,
 119, 125, 145, 153, 157–8,
 161, 163, 189, 191, 192, 198,
 205, 215–16, 226, 251–2,
 262
Wallop Valley 46, 65, 98–9,
 133–4, 142–3, 163, 170, 175,
 188, 195, 216, 240, 253
Walton, Izaak: *The Compleat
 Angler* 49–50
water-forget-me-not (*Myosotis
 scorpioides*) 159, 160
water-mint (*Mentha aquatica*) 158,
 159, 160
watercress (*Rorippa nasturtium-aquaticum*) 158, 159,
 160
water vole, European (*Arvicola

amphibius*) 12, 27, 100, 105,
 106, 139, 140–2, 148, 155, 177,
 183, 220, 241, 2257
willow tree 5, 7–8, 12, 17, 46, 84,
 96–7, 124, 134, 165, 193–5,
 197–8, 213, 256–7
Willow (daughter of Kuschta):
 appearance 118, 154
 companionable relationship
  with Wisp 240, 245–6
 future of 180, 260
 early months as cub 118,
  119
 hunting 125, 128, 131–2,
  146–7, 148, 149, 180, 181,
  183, 184, 185, 211, 214–15,
  217–21, 227–8, 233, 243–5,
  249, 250–1, 252, 256
 Lutran banishment from
  family group, reaction to
  212, 215, 239–40
 mother (Kuschta) abandons
  211, 246, 248, 249–60, 261,
  262
 navigation 241–2
 weight 180
Willow Island 95–7, 101, 112, 134,
 212, 213, 250
winter 27–63, 75, 83, 95, 102,
 105–6, 107, 109, 121–3, 144,
 153–4, 155, 156–8, 160,
 176–86, 187–208, 224, 233,
 246, 248, 253
Wisp (daughter of Kuschta):
 appearance 118
 companionable relationship
  with Willow 240, 245–6
 early months as cub 118,
  119

future of 180, 260
hunting 125, 126, 128, 131–2,
   148, 149, 180–1, 183,
   184–5, 186, 233, 236,
   243–6, 254–5, 259–60
Lutran banishment from

family group, reaction to
   212, 215, 239–40
mother (Kuschta) abandons
   211, 246, 249–60, 261, 262
navigation 241–2
weight 180